Fringed with
Mud and Pearls

FRINGED WITH MUD AND PEARLS

An English Island Odyssey

IAN CROFTON

BIRLINN

First published in 2021 by
Birlinn Limited
West Newington House
10 Newington Road
Edinburgh
EH9 1QS

www.birlinn.co.uk

ISBN 978 1 78027 665 6

British Library Cataloguing in Publication Data
A catalogue record for this book is available from the British Library.

Designed and typeset by Biblichor, Edinburgh
Printed and bound by Clays Ltd, Elcograf S.P.A.

MIX
Paper from
responsible sources
FSC® C018072

What one thinks of any region, while traveling through, is the result of at least three things: what one knows, what one imagines, and how one is disposed.

– Barry Lopez, *Arctic Dreams* (1986)

Contents

Maps

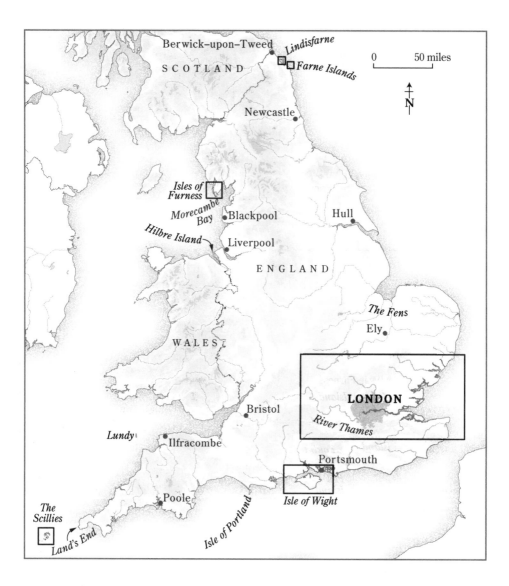

Berwick-upon-Tweed
Lindisfarne
S C O T L A N D
Farne Islands

0   50 miles

N

Newcastle

Isles of
Furness
Morecambe
Bay   Blackpool   Hull
Hilbre Island   Liverpool

E N G L A N D

The Fens
Ely

W A L E S

LONDON
River Thames

Bristol

Lundy   Ilfracombe
Portsmouth

Poole
Isle of Wight
Isle of Portland

The
Scillies
Land's End

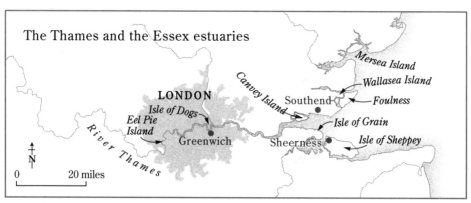

The Thames and the Essex estuaries

Mersea Island
Wallasea Island
LONDON   *Canvey Island*
Isle of Dogs   Southend   *Foulness*
Eel Pie   *Isle of Grain*
Island   Greenwich
River Thames   Sheerness   *Isle of Sheppey*

N

0   20 miles

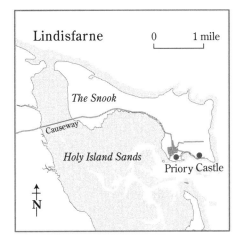

Lindisfarne

The Snook

Causeway

Holy Island Sands

Priory Castle

0   1 mile

N

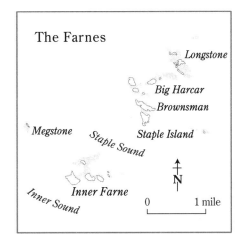

The Farnes

Longstone

Big Harcar

Brownsman

Megstone   Staple Island

Staple Sound

Inner Farne

Inner Sound

N

0   1 mile

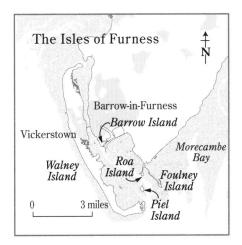

The Isles of Furness

N

Barrow-in-Furness

Barrow Island

Vickerstown

Walney Island

Roa Island

Morecambe Bay

Foulney Island

Piel Island

0   3 miles

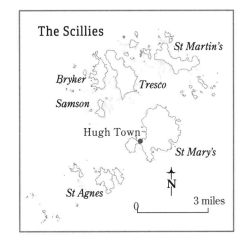

The Scillies

St Martin's

Bryher

Tresco

Samson

Hugh Town

St Mary's

St Agnes

N

0   3 miles

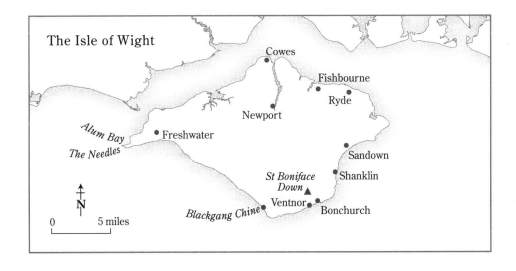

The Isle of Wight

Cowes

Fishbourne

Ryde

Newport

Alum Bay

The Needles

Freshwater

Sandown

Shanklin

St Boniface Down ▲

Ventnor

Blackgang Chine

Bonchurch

N

0   5 miles

Preface

Island-going, like mountaineering, has a large number of
enthusiastic devotees who have really no very clear idea
of why they do it.

– Kenneth Williamson and J. Morton Boyd, *St Kilda Summer*
(1960)

Greece has its sun-soaked Cyclades, its Dodecanese, its elusive
Islands of the Blessed. Scotland has its chilly Northern Isles, its
rain-drenched Hebrides. Wales has Ynys Mon, Skomer, Skokholm
and Bardsey, the Island of 20,000 Saints; Ireland the Arans,
Rathkill and the Skelligs – fangs in the Atlantic where monks
once prayed in beehive huts, sheltering against the elements and
temptation.

And what has England got? The isles of Canvey, Sheppey,
Foulness, Wight and Dogs. Mersea, Wallasea, Two Tree and Rat.
But there are also wilder, rockier places – Lundy, the Scillies, Hilbre,
the Farnes. England is surrounded by a fringe of mud and pearls –
tidal flats and marshes, holiday parks and petrochemical works,
jagged cliffs and silver beaches. Then there's the deserted wartime
forts in the Thames Estuary, oil platforms in the North Sea,
offshore wind turbines, the lost lands of Doggerland and Lyonesse,
the tide-washed sands of Goodwin.

In 2013 I followed England's northern edge, its frontier with
Scotland, for my book *Walking the Border*. That border, like all
borders, is an arbitrary human construct, but following it on
foot revealed many insights into what it is like to live along a

dividing line, however random. It also brought me into contact with how a certain stretch of landscape – chosen as if by a throw of the dice – has been shaped both by nature and by history.

Having completed one small section of a circle of England, my mind turned to the country's other edges, its eastern, southern and western coasts, and, more specifically, those bits off the coasts that have become detached – England's islands. This restriction – that the islands had to be technically part of England – not only ruled out the islands off the Welsh coast. It also ruled out the Channel Islands and the Isle of Man, which are not only not English, they aren't even part of the United Kingdom, merely 'dependencies of the crown'. In this way, I set the rules for myself.

Another rule I soon made was that I didn't need to visit every English island: this was not to be a work of reference or a guidebook. I would make sure I paid heed to the famous ones, but would omit dozens of rocks (so no Goose, no Chick, no Cow and Calf, no Beeny Sisters, no Bull, no Minnows) and countless drifts of shingle, sand or mud (so no Plover Scar or Devil's Bank, no Buxey Sand or Dengie Flat). Also omitted are various islands that are barely known as islands, such as Portsmouth, which occupies virtually the whole of Portsea Island, England's fourth largest island, which in turn is separated from mainland Hampshire by Ports Creek. Portsmouth warrants an entire book to itself. In contrast, its neighbour, Hayling Island, although England's seventh largest island, does not seem to have sufficient history or island identity to warrant even a chapter (although no doubt the inhabitants would beg to differ). It barely receives a mention in J. H. Ingram's *The Islands of England* (1952) – it is, Ingram says, 'one with the surrounding coastline', and lacks 'the authentic flavour'. Hayling is omitted entirely by Donald McCormick in his *Islands of England and Wales* (1974).

I did not have the comprehensive ambition of either of these authors. My idea was to use *some* of England's islands as a range of lenses through which to view the motherland, in all its kaleidoscopic variety. No single lens was ever going to provide a definitive image; nor were a dozen. There is no single truth about England, only a range of viewpoints. I had originally thought of arranging the chapters geographically; in the end, I decided to arrange them in the order of my visits. So it became – as it had been for Odysseus – a hither-and-thither sort of a voyage.

I conceived of this book in the spring of 2014, and that June visited Canvey Island, one of the English islands that is closest to my home in London, just down the Thames Estuary, accessible on a day trip by train. It was an exploratory venture, to see whether even such an apparently mundane (and much mocked) place could yield riches. I found that it could. But I had too much on over the following months to pursue the project further. There were two other books to complete, and various climbing trips scheduled – to Scotland, the Italian Alps, the sandstone towers of the Rheinland-Pfalz. The Scottish trip took me to some islands, including Hoy in the Orkneys to climb its Old Man, and the isolated sea stack of Am Buachaille, a tower of sandstone off the northwest coast of Sutherland. The former is accessible via a couple of ferries, but to get to Am Buachaille you have to swim across a narrow but turbulent channel of the Atlantic, or creep along a rope fixed by the first person over.

Then, in April 2015, my project suffered what might have proved a terminal blow. While rock climbing on Froggatt Edge in the Peak District, I fell thirty feet onto my head. I don't remember anything about it, and have no idea why I fell. I was airlifted to Sheffield Northern General Hospital, where I spent ten days. I'd broken some ribs, chipped a few vertebrae, fractured a shoulder blade and suffered three brain bleeds. These last were subsequently categorised as 'severe traumatic brain injury', and led to double vision, fatigue and what the neuropsychologists called 'executive dysfunction' (which for me meant an inability to organise, combined with confusion and anxiety). Back home in London, my life became filled with endless appointments with specialists in three different hospitals – for bones, eyes and brain; not to mention visits to osteopaths, physiotherapists and various other practitioners. I had neither the time nor the energy – nor indeed the physical and mental capacity – to consider writing another book.

Among the many medical professionals I saw, the most helpful was my neuropsychologist at the National Hospital for Neurology and Neurosurgery, Dr Mary Summers. It was she who encouraged me to resume my island project. She suggested that I try making a trip out of London once a week, talking to people I encountered, and writing up the results. The caveat was that I should not spend more than half an hour at the computer at any one time, and that I needed to take a lie-down rest for one hour every afternoon.

In the event, it was not until early in 2017 that I felt able to continue with my explorations, and to resume writing – at first slowly and painfully, and then with growing confidence. So the book has indeed become an odyssey, a journey of recovery, a difficult return to the place I was before I fell.

~

There are several people I would like to thank for playing a part in the making of this book in addition to Dr Summers. There is Bob Appleyard, who had the misfortune to be climbing with me the day I fell, and was subjected for several minutes to the belief that I was dead (until, apparently, I started twitching and swearing). Two years later Bob and his wife Jan organised my trip to the island of Lundy, and Bob led me up Lundy's sea cliffs via the Devil's Slide, my first multi-pitch rock climb since the accident. I am also indebted to the organisational skills of Alice Harper and Tom Jones, who accompanied me and my wife Sally to the Scilly Isles, the Isle of Portland and the Isle of Wight. On the Isle of Portland we stayed in the delightful cottage of our London neighbours, Richard and Jane Green. Two other neighbours, Dick Sadler and Steve Arthurell, indulgently agreed to join me on a diversion to Two Tree Island as we walked up the Thames Estuary, although in the end I omitted that chapter: Two Tree Island offers but slim pickings; even the two elms that gave it its name are long gone. I am also indebted to Foteini Aravani and William Lowry of the Museum of London, who gave me access to the museum's audio archives relating to the Isle of Dogs; and to Dave Matthews of the Museum of Docklands, who told me about what his father had endured as a docker on the Isle of Dogs, and about the grim institution of gibbeting. Thanks are also due to Chris Davidson and Charles Warren of the Wirral, who enthusiastically filled in the many gaps of my knowledge of Hilbre Island. I would also like to thank Nicola Wheeler and Andy and Rowan Bevan, the present owners of my grandfather's house on the Isle of Wight, who welcomed me when I dropped in on them unannounced. I am, of course, hugely grateful to my editor at Birlinn, Andrew Simmons, who has been helpful and enthusiastic throughout. Finally, I would like to express my gratitude to my wife Sally Mumford, who not only supported me throughout my long recovery, but who also encouraged me in the writing of this book, and cheerfully accompanied me on a number of my island expeditions.

~

This book was conceived, researched and written in a different era to the one in which it is being published. I finished writing it in the late summer of 2019, and it was ready for the press early in 2020. Publication was set for the autumn of that year. But then the world turned upside down, and the schedules of every single publisher around the world were thrown up in the air, together with all other plans, hopes or certainties, whether of individuals, institutions or nations. The vanity of human wishes was thrown into sharp perspective. But my publishers, Birlinn, kept the faith, and their powder dry. So here is the book at last, first frustrated by me falling on my head all those years ago, and then thwarted again by the world falling on *its* head. It will be a long journey of recovery.

January 2021

Between Land and Water

An Introduction

> This fortress built by Nature for herself
> Against infection and the hand of war . . .
> Against the envy of less happier lands.
>
> – William Shakespeare, *Richard II*, II.1

England is troubled by its insularity. Proud of it, prickly with it. It was England Shakespeare was thinking of when he wrote of 'this sceptred isle'. Yet England, 'This precious stone set in the silver sea', is not in itself an island. If you look at the map, you'll see that the Rivers Tweed and Annan almost cut it off from (most of) Scotland. But not quite. Offa's Dyke makes an effort to cut off Wales. The English have a tendency to ignore, or at least disparage, the other countries with which they share the island of Britain. England has become a creature of its own imagination. An island is often more a state of mind than a geographical reality.

After the 2016 EU referendum, England – or at least the Little England of its own devising – started to draw in its skirts. Its islands constituted the hem – not a neatly stitched hem, but a frayed edge. And in these edgelands, these borderlands, I found that the islanders, especially those facing towards Europe, were more inclined than many mainlanders to assert their Englishness, to raise their flags of St George against the dragon of difference, 'against infection and the hand of war . . . against the envy of less happier lands'.

But even the Britain that England so often claims as its own was not always an island. And even today some of the islands round England are not always islands, their land not always land, the water around them not always water . . .

Islands are elemental places, parcels of land circled by the sea. Both these elements, earth and water, are shaped by a third element, air: the wind raises waves, blows sand, erodes stone. Islands formed from igneous rock – the granite of Lundy, the dolerite of the Farnes – also hold a memory of the fourth element, fire, the molten heat from which they were born. Such rocky islands are distinct from the liquid element: firm, hard, immutable as pearls. Others, such as the Isle of Wight, are less distinct, their chalks and clays crumbling constantly into the sea.

Sometimes the boundary between earth and water is even murkier. This is the realm of sand, even more the realm of mud. Sandbanks and mudflats shift, dissolve, re-form. Along soft coasts the sea penetrates the land, washes it away. Elsewhere silt and gravel build up and the sea recedes, sometimes over centuries, sometimes in a sudden storm.

Hard or soft, islands change their size and shape with the twice-daily tide, a rhythm determined by the distant moon. Inflow, outflow, like a body breathing or a jellyfish moving through the water, or a semblance of coition.

The land we know, edged and defined by the sea, is not the land as it once was. Nor is the sea ever the same. In the age of ice there was no North Sea separating Britain from mainland Europe. Millions of cubic miles of water were locked up in the icecaps, leaving large areas of dry land. The North Sea was further north, and in its place was the place called Doggerland. The English Channel was not then a channel at all, nor English, because there was then no such thing as England. Rather it was a gulf, a dead end, an inlet of the Atlantic Ocean, an inlet fed by two great confluent river systems, the ancestors of the Thames and the Rhine, which flowed together to form what has been named the 'Channel River'.

While to the north the ice still lay thick, this low-lying land was tundra. In 1931 a trawler fishing in the North Sea dragged up the point of a reindeer's antler that had been worked to form a barb, suggesting that this was once a Palaeolithic hunting ground. Other tools and weapons have since been recovered from what is now the seabed, together with the remains of extinct megafauna such as mammoths. In those days, should you have been so inclined, you could have walked dryshod from London to Berlin, if those cities had then existed.

As the climate warmed, the ice began to melt and sea levels rose. Doggerland became a more amphibious place, an area of coastal lagoons, mudflats, saltmarshes, inland lakes, rivers and streams. Game, waterfowl and fish were all abundant, providing Mesolithic hunters with rich pickings. The ice-age megafauna, such as mammoths and woolly rhinoceroses, had by this time died out, perhaps due to overhunting.

As the ice melted and sea levels rose, Doggerland began to disappear, and by 8,500 years ago Britain ceased to be a peninsula of Europe. It had become an island. Doggerland – at least vast tracts of it – may have been finally extinguished some 300 years later by a megatsunami, caused by a massive underwater landslide in the North Atlantic, off the edge of Norway's continental shelf. This event is known as the Storegga Slide, *storegga* being Norwegian for 'great edge'.

All that was left of Doggerland was a small island, which may have survived until around 7,000 years ago. A remnant of this island remains in the form of the Dogger Bank, a submarine sandbank some 150 by 60 miles in size between northeast England and Jutland. The Dogger Bank forms one of the shallowest areas of the North Sea, being in places only eight fathoms (not quite fifty feet) below the surface. It has traditionally been a rich fishing ground, and is named after the old Dutch *doggers*, a type of fishing boat specialising in catching cod. The Dogger Bank in turn gave its name to the lost land of Doggerland.

~

The names of many of the islands around Britain end in -*ay* (e.g. Scalpay, Pabbay, Mingulay), or -*ey* (Anglesey, Canvey, Walney, Sheppey), or -*ea* (Mersea, Wallasea), or -*y* (Lundy). In some cases (especially in the north and west, areas of stronger Viking influence), this suffix is from Old Norse *ey*, 'island', cognate with the Old English (Anglo-Saxon) word *eg* or *ig*, also meaning 'island'; it is the latter that has provided the suffix for many islands in the south and east, whose names are often tautologous – the Isle of Sheppey, for example, is 'isle of the island where sheep are kept'.

Most places round England that have 'island' in their name are actual islands (or once were, in the case, for example, of Barrow Island; mention should also be made of Thorney Island, a former island in the Thames where the Palace of Westminster now sits), but the same cannot be said for all the places that have 'isle' as part of their name. Many such areas – including the Isles of Ely, Athelney, Axholme, Oxney, Purbeck, Grain, Thanet and Dogs – are part of the mainland. In some of these cases, such as Ely, Athelney and Oxney, the Old English -*y* / -*ey* element denotes 'dry ground in marshland'. The 'Isle of' element in the non-insular areas mentioned above similarly denotes a place surrounded by wetland, terrain intermediate between open water and dry land. Or 'Isle' may (in the cases of Purbeck, Grain, Thanet and Dogs) indicate a peninsula, surrounded on two or three sides by tidal water or open sea.

The inland Isle of Ely ('place where eels are found') is the raised area of Kimmeridge clay on which the cathedral city of Ely was built. It is the highest area in the Fens of East Anglia. It was for long surrounded by marshes and meres, and could only be accessed by boat or causeway. After the Norman invasion, the Isle of Ely became the base of an anti-Norman rebellion led by

Hereward the Wake. Legend has it that a Norman knight bribed the monks to show him and his men a safe passage across the marshes, enabling them to root out and crush the rebels. Nevertheless, the Isle of Ely remained in isolation for many more centuries, until the drainage of the Fens began in the early seventeenth century with the construction of a system of canals by Dutch engineers. This met with covert resistance by many Fenlanders, whose traditional way of life, based on fishing and wildfowling, came under threat. Attempts to sabotage dykes, ditches and sluices were not uncommon, but by the end of the eighteenth century the Isle of Ely was surrounded not by marshes and meres but by well-drained farmland.

The word *island* itself is from the Old English *eg* or *ig*, with the addition of Old English *land*, which meant then what it means today – that part of the surface of the earth that is solid, rather than water. The 's' was introduced into the word from *isle*, which entered the English language in the thirteenth century from the Old French *isle*, itself derived from the Latin *insula*, 'island'. *Insula* gave rise to the Latin *insulatus*, 'made into an island', which in turn is the origin of the modern English words *insulated* and *isolated*. The similar word *insular* is from Late Latin *insularis*, also derived from *insula*.

~

When we think of an island, we think of isolation, a place cut off from the mainland, and the mainstream of life. It is this intrinsic quality that has led islands to become places of refuge, of sanctuary, even of holiness. For the same reason, islands have also been chosen as places to build prisons, dump rubbish, bury the dead or locate secretive military installations, out of sight and out of mind of the majority of the population. In contrast, people on the mainland often fantasize about islands as miniature paradises, unspoilt Edens, where they imagine themselves leading untroubled, innocent lives. Such fantasies may also lead to dreams of power, of ruling over a small personal kingdom, free from outside interference, rather in the manner of Robinson Crusoe, that prototypical colonialist, on his desert island.

In the natural world, islands play a special role. It was on the Galapagos Islands that Charles Darwin noticed that there was a range of different finches each occupying a different ecological niche. He concluded that a handful of ancestral finches must have been blown there across the sea from the South American mainland, and, finding no competing creatures, had evolved into a number of different species with diverse adaptations to take advantage of the range of available food sources. The absence of predators on many islands around the world allowed some of the resident birds – from the New Zealand kiwi to the Mauritian dodo – to dispense with the ability to fly. But this in turn made them vulnerable to new arrivals: European mammals such as stoats in the case of the kiwi; hungry sailors in the case of the dodo. A number of English islands provided refuges to burrow-nesting birds, such as puffins and Manx shearwaters. Those on Lundy were almost wiped out when brown rats found their way onto the island from passing ships: a burrow can provide protection from a large airborne predator such as a black-backed gull, but it won't stop a rat. After a campaign lasting many years, rats have now been eliminated from Lundy; there has been a similar success on St Agnes in the Scilly Isles.

Islands can also provide the necessary isolation for indigenous species or subspecies to evolve, cut off from the larger gene pool of the mainland. Sometimes these are either larger or smaller than their mainland cousins: examples include the extinct New Zealand moa, a flightless bird that grew up to twelve feet high; and *Homo floresiensis*, a small species of human, under four feet tall, that lived on the Indonesian island of Flores up to around 50,000 years ago. Far to the west of the Outer Hebrides, St Kilda has its own subspecies of wood mouse, the St Kilda field mouse, which is twice as heavy as its mainland counterpart; closer to England, Walney Island has its own variety of cranesbill or wild geranium, while the Isle of Portland has its own subspecies of rock sea lavender, *Limonium recurvum portlandicum*. Lundy boasts the Lundy cabbage, on which in turn depend two endemic insect species, all of which are threatened by feral goats imported onto the island in the last century. Introduced plant species can also play havoc, a notable example being the invasive mauve-flowered *Rhododendron ponticum*, now eliminated from Lundy after many years' effort, but still found rampaging in many other places, including Tresco in the Scilly Isles.

As well as providing refuges for wildlife, islands can be sanctuaries for gods and their followers. In Greek myth, Zeus, the king of the gods, was born on the island of Crete, in a cave in the mountains, and throughout his infancy he was tended here by a goat, to keep him safe from his vengeful father Cronus. Two other gods, Apollo and Artemis, were said to have been born on the island of Delos. It was on another Greek island, Patmos, that John, the author of the Book of Revelation, experienced his vision of the Apocalypse, which forms the last book of the New Testament. John may have been exiled to Patmos by the Roman authorities for his Christian beliefs, but it is not clear whether being stranded on an island played a part in his visionary experience.

Anchorites (religious hermits) and monastic communities have often favoured islands, just as they have sought out other isolated places, such as caves, deserts and mountains. An early Buddhist text from the second century BC adjures monks to 'live as islands, unto yourselves, being your own refuge, with no one else as your refuge'. (Interestingly, some translations prefer 'lamps' to 'islands'.) Remoteness from the pleasures and distractions of the everyday world helps the pursuit of self-denial, and provides the solitude necessary for contemplation. One of the starkest examples of such a willingness to abandon the world is found high on Skellig Michael, a jagged rock some seven miles off the coast of southwest Ireland. More than 500 feet up the rock are a number of beehive huts, simple stone shelters inhabited from perhaps as early as the sixth century by a handful of monks – no more than a dozen at any one time. Here they would have been surrounded only by gannets and God, subsisting on seabirds and their eggs, fish and shellfish. Around the same time, the Irish monk St Columba arrived on the small Hebridean island of Iona, where he founded the monastery that was to become the cradle of Christianity in Scotland, and an important place of pilgrimage. Also in the sixth century, St Cadfan established a monastery on Bardsey Island, off the Llŷn Peninsula, the earliest Christian foundation in Wales. By the later Middle Ages, three pilgrimages to Bardsey were regarded as the equivalent of one to Rome.

England, too, has its holy islands. One of the most spectacular is St Michael's Mount, a small, rocky, tidal island near Penzance in Cornwall, named after the same dragon-slaying archangel as Skellig Michael. St Michael's Mount was the site of a monastery from possibly as early as the eighth century. In the later Middle Ages it was linked to its Norman namesake, Mont St Michel, another rocky tidal island with a monastery, and attracted many pilgrims. But the most notable of England's holy islands is the Holy Island of Lindisfarne, where St Aidan, a monk from Iona, founded a monastery in 634, and where St Cuthbert became prior some thirty years later. Cuthbert's role in evangelising the north of England, and the many miracles attributed to him, helped turn Lindisfarne into a great pilgrimage destination. It still is, and many religious retreats are held on the island. Similarly, today, some islands have been used as retreats for the purpose of rehabilitation and detoxification. Those wishing to stop smoking have been known to strand themselves on Lundy for a week, having left their cigarettes behind. Osea Island in the Blackwater Estuary was formerly home to a centre for those suffering from addiction problems; among its clients was Amy Winehouse.

Perhaps not unconnected to their spiritual significance, islands have become homes for the dead. There are said to be more prehistoric graves on the Scilly Isles than in the whole of mainland Cornwall, leading to the suggestion (unsubstantiated) that bodies were actually brought here for burial or cremation.

Bardsey itself was called the Island of 20,000 Saints because of the numbers of monks buried there. Eilean Munde, a small island in Loch Leven near Glencoe, was used as a graveyard by three local clans, the Camerons of Callart, the Macdonalds of Glencoe and the Stewarts of Ballachullish, all of whom would share the cost of maintenance even when they were fighting each other. Further afield, Isola di San Michele, a small island in the Venetian Lagoon, was designated in 1807 as a place exclusively for the interment of the dead, it being decreed that burial on the mainland or on the other islands of Venice was unsanitary. Closer to home, Canvey Island has become the final resting place of 3,300 skeletons uncovered during work on the twenty-first-century Crossrail project; the London cemeteries were all full.

Islands can also be more casual dumping grounds for corpses. Deadman's Island, a patch of salt marsh off the Isle of Sheppey, was used as a depository for the corpses of convicts who died on the prison hulks moored nearby. Excavations on tiny, tidal Burrow Island in Portsmouth Harbour recently revealed a number of human skeletons, suggesting that it had been used for a similar purpose. This may be why the place is also known as Rat Island.

Islands are sometimes repositories of non-human waste. Canvey Island and neighbouring Two Tree Island had large landfill sites, and there was a sandbank in the Severn Estuary, where the Royal Edward Dock in Avonmouth is now located, that was known as Dungball, owing to the quantities of rubbish that were once dumped there.

The dead never escape off their islands. Neither do some of the living. In the case of the title character in William Golding's novel *Pincher Martin*, shipwrecked on a small rock somewhere in the Atlantic after his destroyer has been sunk, it is not altogether clear into which category he falls. That, perhaps, is the purpose of those who force others into exile on remote islands – that their victims may experience a living death. In 2 BC the Emperor Augustus had his only daughter, Julia, arrested for treason and adultery. She was alleged to have had affairs with a number of men while married to Augustus's stepson (and successor) Tiberius, and to be plotting against her father. Rather than having her executed, Augustus confined his daughter to the tiny island of Pandateria, with an area of less than two-thirds of a square mile, and situated twenty-five miles off the coast of mainland Italy. Julia was, Augustus decreed, to be deprived the taste of wine, and the sight of any man. Another, larger, island off the coast of Italy, Elba, was the location of Napoleon's first exile; his second, final, exile was to St Helena, 1,200 miles from the African coast and twice as far from Brazil, making it one of the most remote islands in the world.

There is more than a strain of vindictiveness in these exiles, a strain that can also be detected in the decision of Napoleon's nephew, Napoleon III, to dispatch his political opponents to the Îles de Salut, the Islands of Salvation, off the coast of French Guiana. The smallest of these sweltering, disease-ridden islets was designated for the punishment of those guilty of treason and espionage. It became known as Devil's Island; and Alfred Dreyfus, unjustly convicted by an anti-Semitic conspiracy of betraying French military secrets to the Germans, became its most notable inmate. Similarly, from 1961 the apartheid regime in South Africa used Robben Island off the Cape coast to imprison its political opponents, most notably Nelson Mandela.

Institutional incarceration of common criminals is often more casual in its cruelty. Islands are favoured destinations for building large-scale prisons, most infamously Alcatraz in San Francisco Bay, from which, it was said, no man ever escaped alive. Several island prisons are still in operation off the shores of England, including those on the Isle of Sheppey, the Isle of Wight and the Isle of Portland. The inmates of the original 1848 prison on Portland were set to quarrying stone to make the breakwaters that still shelter Portland Harbour. A similar plan had been devised for the Scilly Isles earlier in the nineteenth century. There had long been talk about building a series of breakwaters between the islands to provide a sheltered anchorage for the Royal Navy, closer to the strategic Western Approaches than established bases further up the English Channel. To this end, it was proposed to establish a convict settlement on the Scillies, and make the prisoners break up the rocks of the islands themselves and throw them into the sea. The cost, however, was enormous, and the Admiralty sat on the plans for some decades. In the end, the advent of steam-powered vessels obviated the need for such a base.

There had been an earlier – unofficial – convict settlement on an English island. In the 1740s a local entrepreneur, MP, shipowner and smuggler called Thomas Benson, who had acquired a lease on the island of Lundy, made contracts with the government to transport convicts in his ships to Virginia and Maryland. But rather than taking them to the New World, he landed them on Lundy, where he used them as slave labour to work on his own projects.

Benson was not the first, nor the last, to regard 'his' island as a realm apart, a personal fiefdom, to do with as he pleased. Purchasers of other islands around England's shores have often shared similar delusions. After Augustus Smith acquired the lease of the entire Isles of Scilly from the Duchy of Cornwall in 1834, he styled himself 'Lord Proprietor of the Scilly Islands', and went on to rule as a not-always benign autocrat. Brownsea Island in Poole Harbour – the location of Robert Baden-Powell's first ever Scout camp in 1907 – was purchased in 1927 by the wealthy but reclusive Mary Bonham-Christie. She proceeded to evict the island's 200 inhabitants. After a catastrophic fire in 1934 (which she blamed on the Boy Scouts, who had continued to hold camps there), she banned all visitors from the island, employing security guards to enforce her order. After her death in 1961, the island passed to the Treasury in lieu of death duties, and was subsequently transferred to the National Trust.

A similarly happy fate eventually befell Lundy. In the nineteenth century it was purchased by William Hudson Heaven, who had made his money from the family slave plantations in Jamaica.

Heaven refused to acknowledge any authority from the main-land, or to allow any official to land on what became known as 'the kingdom of Heaven'. This attitude was maintained by Martin Coles Harman, a City financier who acquired Lundy in 1925, and declared that the island was 'a little Kingdom in the British Empire, but out of England'. Harman reigned as a benign despot; among other decrees, he banned his 'subjects' from submitting tax returns to HM Government, on pain of banish-ment. Harman's successors could not afford to maintain their little kingdom, however, and in 1968 sold Lundy to the National Trust.

Perhaps the starkest example of the vanity of such would-be monarchs is the story of Roughs Tower, a Second World War anti-aircraft fort that stands in the North Sea half a dozen miles off the Suffolk coast. The concrete and steel tower, originally commissioned as HM Fort Roughs, was built in a dry dock in Gravesend and installed in 1942 on Rough Sands, a sandbank guarding the approach to the port of Harwich. Naval personnel were eventually withdrawn from Roughs Tower in 1956, but ten years later it was occupied by a couple of pirate radio operators, Paddy Roy Bates of Radio Essex, and Ronan O'Rahilly of Radio Caroline. The two shortly fell out, and Bates, who had been an army major during the war, seized the tower for himself. O'Rahilly tried to retake it, but was fought off with guns and petrol bombs. The British authorities arrested Bates and his son Michael on fire-arms charges, but the courts threw out the case, as Roughs Tower was outside UK territorial waters (which were then limited to three nautical miles offshore). Thus encouraged, Bates went on to declare Roughs Tower to be the Principality of Sealand, with its own flag, constitution, national anthem and passports. He himself adopted the title Prince Roy of Sealand. Uneasy lies the head that wears the crown, however. In August 1978 Alexander Achenbach, a German businessman who claimed to be prime minister of Sealand, hired a team of Dutch and German merce-naries to storm Roughs Tower while Prince Roy was on the mainland. The mercenaries, who deployed jet skis, helicopters and speedboats, were initially successful, taking Bates's son Michael hostage. However, Michael had hidden a cache of weapons in the tower and was able to retake Sealand from the rebels. Achenbach, who held a Sealand passport, was held and charged with treason. The foreign ministries of the Netherlands, Austria and Germany asked the UK to intercede, but the latter declined, citing the earlier court ruling. A German diplomat was then dispatched to Roughs Tower, and after weeks of negotiation Achenbach was released. The former prime minister went on to set up a government in exile. In 1987 the UK extended its territorial waters to twelve nautical miles offshore, and the United Nations Convention on the Law of the Sea, signed by 165 countries and in force since 1994, states: 'Artificial islands, installations, and structures do not possess the status of islands. They have no territorial sea of their own, and their presence does not affect the delimitation of the territorial sea, the exclusive economic zone or the continental shelf.' After Bates retired to the mainland, Michael took over as prince regent. Bates himself died in 2012.

Roughs Tower was one of a number of defensive forts built in the outer Thames Estuary and off the coasts of Essex and Suffolk during the Second World War, primarily as anti-aircraft platforms. In the previous century, another series of defensive forts had been constructed round the coasts of England, at great expense. They were mostly built in the decade after 1860, strongly promoted by the prime minister, Lord Palmerston. But by the time they were completed the perceived threat from the French had disappeared following their defeat by the Prussians in 1870–1. By that time developments in artillery technology had anyway rendered 'Palmerston's Follies' obsolete. Some of the Follies – such as No Man's Land Fort and three others in the Solent – are artificial islands like Roughs Tower. They never saw action, and some of them are now luxury hotels, equipped with swimming pools and helipads. (There are other Palmerston Forts on the Isle of Wight and the Isle of Portland.) Other examples of small artificial islands round England's coasts include the great break-waters sheltering Portland Harbour, offshore wind farms (some of the biggest arrays are off Walney Island, near Barrow-in-Furness) and the natural gas platforms off Great Yarmouth.

Some of England's non-artificial islands continue to play significant military roles. Thorney Island in Chichester Harbour, separated from the mainland by little more than a couple of ditches, is largely occupied by an army base, and access is restricted to the coastal path. Even more restricted is Foulness Island off the Essex coast, where the British School of Gunnery was opened in 1848. Foulness is now an MoD weapons-testing installation, run by QinetiQ, a private company. The public can only visit the island on a single day every month, and must stay on a single road. The only alternative is to walk the Broomway, an ancient right of way across Maplin Sands. This is only possible when the tide is out, and should only be attempted in the company of a local guide. My own attempts to visit the island were thwarted. In March 2017 I had booked to go on a small boat to look at Foulness from the water, but the weather turned as my bus approached Burnham-on-Crouch, and the sailing was cancelled on the advice of the coastguard. Then that summer I booked to walk the Broomway with a local guide, but that too was cancelled; as the UK government declared a yet higher level of security alert, the MoD closed off the Broomway for several months for exercises. The nearest I got to Foulness was on neighbouring Wallasea Island, from where I witnessed a number of explosions. I was startled; the wildfowl of Wallasea remained indifferent.

By their very nature, islands are in the front line as far as seaborne invasions are concerned. The first large-scale Viking raid on England targeted Lindisfarne in AD 793. The French landed on the Isle of Wight in 1545, and the Dutch on the Isle of Sheppey in 1667. There are many reminders of the Second World War on the islands of the Thames and the Essex estuaries, mostly in the form of pillboxes. The Isle of Wight boasts, in addition to a couple of

Palmerston Forts, an abandoned radar station used during the Second World War, together with a Cold War nuclear bunker. At low tide off the Isle of Sheppey you can still see the three masts of the SS *Richard Montgomery*, a Liberty ship that sank in the Thames Estuary with a massive cargo of bombs towards the end of the Second World War. If they ever detonate, the result will be the biggest non-nuclear man-made explosion of all time.

That will be an even bigger explosion than the one that nearly destroyed Heligoland, Britain's long-forgotten North Sea island.

Heligoland is a small lump of red rock lying forty or so miles off the coast of northwest Germany, not far from Denmark. Before the seas rose at the end of the last ice age, it would have lain at the eastern side of Doggerland, the low-lying land that once linked England and continental Europe. In 1807, during the Napoleonic Wars, the British seized the island from the Danes and used it as a base for covert operations against Napoleon's continental empire. In 1814 it was formerly ceded to Great Britain. It remained a British colony until 1890, when the newly united Germany, realising the strategic importance of Heligoland for its navy, exchanged various areas of influence in East Africa for the island (in fact two islands; the main island has a low-lying uninhabited neighbour). So for three-quarters of a century Heligoland was part of the British Empire, 'the point at which Great Britain and Germany come most nearly into contact with each other', according to a senior official in the Colonial Office in 1888. It was, he said, 'the only part of the world in which the British government rules an exclusively Teuton though not English-speaking population'. Under British rule Heligoland thrived on a combination of smuggling and tourism, as the island became a refuge for many German radicals and revolutionaries fleeing political oppression on the mainland. The poet Heinrich Heine visited the island on a couple of occasions to escape the censors and the police, although he was not impressed by the island's rulers: every Englishman he met, he wrote, gave off 'a certain gas, the deadly, thick air of boredom'.

As part of the German Empire from 1890, Heligoland was transformed into a heavily fortified naval base, and continued in this role under the Nazis. RAF attacks began in 1944, culminating in two massive waves involving a thousand bombers on 18–19 April 1945. Although most of the inhabitants survived in their underground shelters, the island was rendered uninhabitable, and the entire population was evacuated. With the defeat of Nazi Germany shortly afterwards, the British took over Heligoland once more, and used the empty island as a bombing range. On 18 April 1947 the Royal Navy detonated 6,700 tons of explosives on the island, with the double aim of disposing of surplus munitions left over from the war, and of demolishing the island's fortifications, including its U-boat pens. It was accepted that the island itself might be destroyed. There were scientific reasons, too, and seismologists welcomed the opportunity of measuring the effects of the blast. When the explosion occurred, it was with a force one-third of the atomic bomb that destroyed Hiroshima. A black mushroom cloud rose 12,000 feet into the air. Although the island itself survived, its structure was radically altered. It was not until 1952 that Heligoland was returned to German control, and the islanders were allowed to return.

Today, Heligoland is once more a tourist destination, a particular lure being its tax-free shops. As in its smuggling days, Heligoland thrives on a trade in tobacco and alcohol. Although it is part of the European Union, Heligoland is excluded from both the customs union and the EU VAT area.

Similar anomalies arise with some of the islands closer to England's shores, such as the Isle of Man and the Channel Islands. None of these islands are actually part of England. They are not even part of the UK. They are rather 'crown dependencies'. On the Isle of Man (once Norwegian, then Scottish, before coming under the feudal rule of the English crown) the British monarch (whether king or queen) is Lord of Mann, while on the Channel Islands the monarch (again, whether king or queen) is Duke of Normandy, the Channel Islands being all that is now left of the dukedom once ruled by William the Conqueror. The Isle of Man has never been a part of the European Union, nor have either of the Channel Island crown dependencies, the Bailiwick of Jersey and the Bailiwick of Guernsey (which also includes Alderney, Sark and some smaller islands). All three crown dependencies are domestically self-governing, while the UK government looks after foreign affairs and defence. Sark itself had, until reforms in 2008, a government headed by the hereditary Seigneur (or Dame) of Sark, under the feudal overlordship of the British monarch. There is now a largely elected legislature. Feudal overlordship also persists on the Isles of Scilly, where virtually every freehold is held by the Duchy of Cornwall, the vast money-making estate established to keep successive Princes of Wales in the manner to which they are accustomed.

The Channel Islands and the Isle of Man are just some of the territorial anomalies that surround England. Among the inland, non-maritime islands is one in the Tweed, the river which along some of its length marks the border between England and its northern neighbour. A few miles downstream from Coldstream, the Tweed is for a short while bifurcated by a small, low-lying island, covered in willows and scrub. Half way along the length of this island, the Anglo-Scottish border briefly kinks at right-angles to its normal course, so dividing the island between the two nations. This divided sovereignty is reflected in two different names: the southwestern, English portion is Dreeper Island, while the northeastern, Scottish portion is Kippie Island. No flag flies on either side of the invisible line across this single, small landmass.

So islands can tell us something of the absurdity of borders, the conceit of human claims to ownership, the vanity of dividing a single, small world into patches of sovereignty. These transient scraps of territory have over geological time alternated between water and dry land as sea levels rise and fall, and as the tectonic plates that make up the earth's skin shift and jostle. Everything, they tell us, is in flux. No island, no patch of land, is for ever. Nothing, in the end, endures.

Out of the Estuary, Kicking Arse

Canvey Island

I was born here. So that means I was born below sea level.
This affects the consciousness profoundly.

> – Wilko Johnson, of Canvey Island R&B pub band
> Dr Feelgood, in Julian Temple's documentary
> *Oil City Confidential* (2009)

Most people of a certain generation know Canvey Island not for its largely defunct petrochemical works, but as the home of the early 1970s R&B pub band Dr Feelgood. With their choppy guitar style, their mod suits and their Essex accents, Dr Feelgood were punks before punk.

'There was an underlying sense of violence,' one fan recalls. 'And there was a lot of alcohol.'

Another comments, 'Audiences thought they might have been watching four guys who'd just done a bank job.'

The band were, according to a third fan, 'Out of the Essex Estuary, and kicking arse.'

~

Today the Thames Estuary forms a great seaway into southeast England, a funnel that draws in shipping from all round the world, a funnel that narrows and narrows until the two banks are close enough to be bridged.

Two thousand years ago it was a different matter. Long before the river's bed was dredged to make a channel deep enough for ocean-going ships, long before its shores were defined with sea walls and embankments, the Thames Estuary sprawled far to either side of where it lies today. It was a vast area of marsh and tidal flats threaded with shallow channels, an area that was not quite land, not quite sea, mostly mud. After the Roman legions of Aulus Plautius defeated Caratacos, chief of the Catuvellauni, at the Battle of Medway in AD 43, the beaten Britons made their escape at low tide north from Kent across the Estuary by a series of fords. Their route was probably across the reach of the Thames now known as the Lower Hope, perhaps between Cliffe Fort on the Kent coast and Coalhouse Fort on the Essex shore below Tilbury.

It is impossible to say whether this was a regular crossing. The earliest known bridge across the Thames was built by the Romans themselves, in the settlement they founded and called Londinium. This first London Bridge, like its medieval and Victorian successors, was built at the lowest point on the Thames to be bridged until the opening of Tower Bridge in 1894; this was itself superseded with the opening of the Queen Elizabeth II Bridge far downriver at Dartford in 1991. Besides the bridges, there were – and still are – many ferries to take both people and vehicles across the river. Several tunnels carry out the same function, some for trains, some for road vehicles, some for pedestrians alone.

Dredged, embanked, bridged, tunnelled, the Thames still carves its way through the land, its flow forever seaward, its tides ebbing and flowing with the moon, revealing and concealing broad flats and banks of mud or dirty, soft sand, where earth and water make their twice-daily interchange. Great stretches on either side of the main channel are neither sea nor land, safe neither for ship nor human. If you were to take a boat down the Thames and let the river's flow and the ebbing tide take you seaward, you'd pass through stretch after stretch of water each of which has been given a name by mariners and chart-makers, names such as Gallions Reach, Halfway Reach, Fiddler's Reach, Northfleet Hope, the Lower Hope, Sea Reach. At Sea Reach, between St Mary's Marshes on the Kent coast and Deadman's Point on the Essex shore, you'd be floating past a low-lying tract of land recovered from the sea by human hand, just separated from the mainland by the narrow gulch of Benfleet Creek. This is Canvey Island.

William Camden, writing in the later sixteenth century, describes how Canvey was in his day 'so low that often times it is quite overflowen, all save hillocks cast up, upon which the sheep have a place of safe refuge'. Although over the centuries various flood defences had been built around Canvey, the island was still, at least on Camden's evidence, frequently flooded. But in 1622 the various landowners of Canvey got together and hatched a scheme to provide the place with more effective and permanent protection from the sea. To this end they commissioned several hundred Dutch engineers practised in the arts of land reclamation to build dykes and begin the work of draining the island, round which they then built a wall. Many of the Dutch engineers were paid with grants of land on Canvey, where they raised cattle. To this day many roads on Canvey have names of Dutch origin – Kamerwyk, Dovervelt, Paarl, Roosevel, Waarden. (This Dutchification may have been down to a property developer called Frederick Hester, who at the turn of the last century made the first attempt to turn Canvey into a holiday resort for Londoners, promoting the place as 'Ye Old Dutch Island'. Hester and his scheme went bust in 1905.)

~

I left London for my first English island from Fenchurch Street, the only mainline station in the capital not connected to the Underground. So it is something of an island itself. To get to Fenchurch Street you have to get off the District and Circle at Tower Hill, at the back of the Tower of London. Many think of the Tower as the heart of England, which I suppose it has been for nearly a thousand years, even though it was built on the orders of the French-speaking boss of a gang of greedy and violent Norman adventurers. The intention was to impress the English that the new rulers meant business; anybody stepping out of line would be treated with the utmost brutality.

On the other side of the Thames, beyond Tower Bridge, you see another, more recent symbol of power, the Shard. It's like a broken bottle grasped in a thug's fist, all set for a fight. This most dominant component of the London skyline was designed by an Italian, and is largely owned by the State of Qatar.

To get to Fenchurch Street Station from Tower Hill I skirted Trinity Square, dominated by the former headquarters of the Port of London Authority, an Edwardian-style extravaganza adorned with a statue of the sea god Neptune, together with a clutch of Nereids and Tritons. More modestly tucked to one side is the home of the Corporation of Trinity House of Deptford Strond, the organisation that looks after the lighthouses of England and Wales. It all seemed a good omen for the start of my island journey, and I hoped for safe passage as I crossed the waters. All these buildings and institutions also reminded me that being an island does not necessarily cut a place off from the rest of the world. For centuries – with unpaved roads baked into impassable ruts in summer, churned into quagmires in winter – it was often easier and quicker to travel by sea than by land.

Before the railways and the advent of passable roads the only way to go down the Thames from London into the Estuary and the open sea was by boat. Even then it was often not easy to determine whether one was travelling over earth or through water, the two so often being combined into that mongrel element, mud. As fog, that other mongrel element, dominates the opening of Dickens's *Bleak House*, so mud is a recurrent note in a passage towards the end of *Great Expectations* in which Pip attempts to smuggle Magwitch, the escaped convict, onto a paddle steamer to Hamburg, somewhere along 'those long reaches below Gravesend, between Kent and Essex, where the river is broad and solitary . . .' Both fog and mud are emanations of the Thames, and both are symbols of stasis, of spiritual paralysis.

> It was like my own marsh country, flat and monotonous, and with a dim horizon; while the winding river turned and turned, and the great floating buoys upon it turned and turned, and everything else seemed stranded and still . . . some ballast-lighters, shaped like a child's first rude imitation of a boat, lay low in the mud; and a little squat shoal-lighthouse on open piles, stood crippled in the mud on stilts and crutches; and slimy stakes stuck out of the mud, and slimy stones stuck out of the mud, and red landmarks and tidemarks stuck out of the mud, and an old landing-stage and an old roofless building slipped into the mud, and all about us was stagnation and mud.

At last they make landfall, and spot the light of a public house, 'a dirty place enough, and I dare say not unknown to smuggling adventurers'. This pub has long been identified as the Lobster Smack on Canvey Island. Apart from the landlord and his wife, the only other human they encounter is a small man, a person both 'slimy and smeary'. This amphibian-like creature specialises in recovering the corpses of the drowned from the water in order to salvage their clothing. His booty is usually found to be 'in various stages of decay'.

It's easier to reach Canvey Island today. If you alight from the Fenchurch–Southend train at South Benfleet, a low, flat concrete bridge will take you over the mud and brine of Benfleet Creek to the island itself. The greatest difficulty you will have is crossing the very busy main road. I had to wait several minutes for a gap in the traffic, for this is only one of two roads accessing the island, whose population is now nearly 40,000.

Benfleet Creek is home to a few moored boats, their rotting hulks flat and motionless on the water. A thousand or more years ago it would have presented a different picture. The place that the Anglo-Saxons called Beamfleote was the site of a fort built by the Danish commander Haesten, close to the creek where his fleet could take shelter. The estuary that provided Pip and Magwitch with the promise of escape also provided a seaway for Viking invasion fleets. Hitherto, the Vikings had taken advantage of divisions among the Anglo-Saxon kingdoms into which England was then divided. But by 893 the West Saxons under Alfred had formed some kind of entente with the Mercians, whose kingdom was centred in the west Midlands. At this date, London had a Mercian garrison, and it was troops from this garrison who took advantage of Haesten's absence when he was off raiding elsewhere. They attacked Benfleet, captured Haesten's family, and burnt his fleet. When the railway line was being built in the 1850s, the navvies dug up large numbers of human bones and charred timbers. A local legend claims that on Canvey Point, on the eastern tip of the island, the ghost of a Viking can sometimes be seen, scanning the horizon for a ship to take him home.

There were no ghosts in evidence as I crossed over to Canvey Island, only the white stars of stonecrop on the banks of the creek, and the noise of the traffic on Ferry Road. This name commemorates the fact that, before the first bridge was built in 1931, people had to make their way across Benfleet Creek either by a series of stepping stones, or, at high tide, by a rowing boat.

On the right-hand side of the road lies a maze of quiet subsidiary creeks cutting through the pastureland that makes up the western third of the island. The tide was full, and the creeks were crammed with old barges and pleasure cruisers, moored alongside wooden walkways that seemed to have grown out of the mud. Some of the walkways were piled high with rubbish – broken slats, rusted gas cylinders, bits of pallet and carpet, plastic sheeting. Many of the boats themselves were half derelict, their flaking paintwork grubby with algae. Some sat in the water tethered by weed-draped hawsers to buoys and posts, others were drawn up onto the marshland grass, others tilted on mudbanks. It would not be many years, I thought, until they were entirely rotted and reabsorbed into the watery landscape.

On the eastern side of the main road a path led through meadow-
land alongside a low strip of concrete, perhaps six inches high by
a foot wide, zigzagging amongst grasses and mallow and hawk-
weed. Then the strip veered off suddenly, climbed a low bank,
and at the top of it joined a concrete wall, perhaps a foot high
and a foot wide. The path continued alongside this into the
distance. The ground on the other side of the wall, stretching
north towards Benfleet Creek, was lower. Reed beds and stands
of irises were interspersed with clumps of hawthorn, elder and
wild rose. Beyond, I could make out the masts of yachts at anchor.

There was a woman on her own walking a spaniel. I engaged
her tentatively in conversation. After some dog-related small talk,
I asked her if she felt like she lived on an island. Not really, she
said. With only two bridges to the island, didn't she sometimes
feel cut off, I asked. She said she didn't. I mentioned the fact that
much of Canvey was below sea level, and, what was more, downriver of the Thames Barrier, intended to protect London from the
ravages of the North Sea. The Barrier, between Greenwich and
Woolwich, does nothing to protect the less-populated low-lying
land on either side of the lower reaches of the Thames Estuary.
These less densely populated areas – though no one has ever
admitted to this – are presumably expected to absorb much of the
initial impact of any storm surge, so helping to protect London.
The previous winter the Barrier had been closed some fifty times,
while in most years the total number of closures was usually
about half a dozen. How did she sleep at night, I asked. She
shrugged. 'We've got this,' she said, and patted the concrete wall
beside the path.

And so we got onto the subject of the great North Sea flood
which on the night of 31 January–1 February 1953 overwhelmed
many low-lying parts of eastern England, together with a large
area of the Netherlands. A combination of strong winds, low
atmospheric pressure and a high spring tide built up into a mighty
storm surge that swept down through the narrowing North Sea,
gathering intensity as it forged southward, inundating vast tracts

of land under many feet of water. Over 1,800 people died in the Netherlands, where one-fifth of the land is below sea level. In England more than 300 died, fifty-eight of them on Canvey Island.

In the Netherlands they call it the *Watersnoodramp* ('water emergency disaster'). The dykes that had kept so many parts of the country dry for centuries failed in the face of the combination of a northwesterly storm and an exceptional spring tide. The Dutch engineering on Canvey that had reclaimed much of the island from the sea also proved unable to withstand the forces of nature. One of the few buildings to escape damage was the Red Cow pub, built on a slight eminence. After the flood it was renamed, with a certain dark humour, the King Canute. It has now closed down.

I said I'd heard that Canvey had long been a holiday destination for the East End of London. People came from even further away than that, she said. But not any longer. Now the caravan parks had been turned into accommodation for what she called 'benefit people'. I asked whether that caused problems. 'It is what it is,' she said. 'You get it everywhere round here.' I said it couldn't be much fun living in a caravan through the winter. She agreed. 'They're only meant to stay for six months at a time, but . . . ' She left the sentence unfinished.

As she turned for home I carried on along the path beside the concrete sea wall that rings Canvey all the way round. Below me, on the inland side, was the island's golf course, fifteen feet below sea level.

Further on I met three middle-aged cyclists, two men and a woman, all muscled, all in t-shirts and shorts. It turned out they were native Canvey Islanders. They told me they didn't feel isolated, even though Canvey was an island, because these days the place was so built up. They said they cycled round the coast to escape all the new-build that had sprung up since the 1980s, housing thousands of incomers from London. 'As the ethnic minorities moved into the East End,' one of them told me, 'so the English people moved out.' He said that the ways of the incoming kids were rubbing off on the local youth. The result was a sharp rise in crime.

It was difficult to imagine such social problems, standing as we were among the wildflowers next to empty saltmarshes under a huge blue sky, while to the north rose the low hills, fields and woods of the Essex shore, looking much as they must have done for hundreds of years.

'I love it here,' said one of the men.

'I don't like where it's built up,' said the woman. 'They're just building more and more. There's too many people on the island.'

We turned to the 1953 flood. One of the men told me how people first took refuge in their attics, then had to break through the roof tiles to escape the waters. Most of the houses were bungalows. He said that from the heights of Benfleet all you could see of Canvey were the rooftops. The sea defence that had been breached was no more than a length of corrugated steel driven into the ground. At the time, his father had worked at a coal depot, and when the waters rose he'd gone down to the depot and joined some other men in the coal lorry, driving around above the water level rescuing people where they could.

Although none of the three cyclists would have been alive in 1953, the folk memory of the disaster remained strong, encroachment by the sea a constant threat. They remembered that as kids they'd played in the sand dunes at one end of the island. 'The dunes aren't here no more,' one of the men told me. 'The sea's washed a lot of it away.'

Perhaps the memory of what used to be was too painful to confront, because the other man brusquely announced he had to get going. And so they went their different ways, and I mine, leaving me to the song of the larks, the mournful cries of the oystercatchers, the sighing of the wind.

As I walked east the sea wall got higher, and the land on the inland side lower. At the same time, the tide was draining out of the creeks, leaving a thin skim of water over the mud, while narrow rivulets meandered down the centre. Here and there a gull pecked at whatever the sea had left behind.

As I approached the more built-up areas of Canvey, I could make out grass-covered ramparts surrounding grass-filled hollows: the remains of the old Dutch dykes and polders. There was also evidence of more modern engineering efforts to keep nature's excesses at bay, in the form of sluices and pumps. Behind the defences lay the sprawling but densely packed suburbia that now occupies more than half of the island. In places, concrete-lined drainage ditches curved between garden fences. Many of the villas had two storeys, the upper floor at the height of the sea wall. White 4×4s and new black vans with tinted windows were parked in the driveways. Some of the villas had ship-like balcony rails, others mullioned windows, while others boasted pink stucco archways in the Italian manner. This being a World Cup summer, many Crosses of St George were in evidence. There was a general air of defiance and pride. On the other side of the sea wall, the picture was different. In a narrow sub-creek the out-going tide sluiced past the bare skeleton of a skiff, its cracked hull sunk deep in the mud, loaded with a growth of fresh weed, ringed by marsh samphire.

I wondered who lived in these executive-style villas. Then I met Barry's owner. Barry was a mixture of Jack Russell and chihua-hua. 'It's wot you call a mongrel,' Barry's owner told me. Actually, the dog was his son's, but since he was retired he had the job of walking Barry. When he worked, he used to drive the A13 every day to Canning Town. I asked him what he did there. 'Concierge,' he replied. 'I used to wear a bowler hat. Booted and suited.' Had he worked in some mansion block, I wondered. 'Nah, Excel, exhibition venue,' he said. 'I was their man at the front.' He was more relaxed now he was retired, his belly straining against a pale blue shirt. The shirt wasn't tucked in, and the buttons were done up unevenly and intermittently, exposing an area of lobster-red chest. Like the two male cyclists I'd met earlier, he had a full, flushed face and short-cropped hair.

I said my farewells and strode on. Once I was alone again, a pair of goldfinches flew in over the saltmarsh and perched on the edge of the sea wall. In a while the executive villas gave way to pinched terraces of brick-built two-storey council houses. The division was marked by a tall, impenetrable hedge of Leyland cypress. In turn the council housing gave way to an area of single-storey prefabs built out of concrete slabs, widely spaced among short-cut lawns. Between the coastal path and the prefabs stood a ten-foot fence of sharp-tipped steel posts, and a stern sign declaring

PRIVATE
RESIDENTS
ONLY

On the other side of the fence and the sign, a row of steps led downward into the private, sub-sea-level domain of the prefab-dwellers.

Beyond the prefabs and the fence that ringed them rises the low eminence of Canvey Heights. The OS map doesn't mark a single contour or spot height, but Barry's walker had told me that Canvey Heights is the highest point on the island. Like much of Canvey, the Heights are man-made. In this case the mound had originally been the island's dump. It had started as landfill, but once the hole had been filled they kept on piling up the rubbish. It's now covered in grass and bushes, and has been designated a Country Park. As I skirted its edge I found cow parsley, teasel, fennel and a wild white rose, together with pink, purple and yellow vetch, and the fresh green flames of dock. Far on the distant eastern horizon I could just make out Southend Pier. At first I thought the building I was looking at, standing in the middle of the widening Estuary, must have been a defensive fort from the Second World War. Then I realised that it was the Old Pier Head, located more than a mile out from the shore at the end of the world's longest pleasure pier.

My path now took me back west along Smallings Creek, through a flat, well-trodden area between low bushes. The ground was covered with carpet tiles and empty cans. There was evidence that cars came here, perhaps so driver and passenger could indulge in otherwise forbidden fumblings under cover of darkness. Or perhaps it was a haunt for doggers, those who enjoy an audience while engaging in sexual activity. But there seemed little evidence of pleasure, not even a used condom.

Circling round Smallings Creek I came to what signs variously called West Creek Moorings, Oyster Creek Moorings or Smallgains Marina. There was barely a drop of water in the creek, though I could hear the last of the tide trickling away. A network of wooden walkways strode high across the mud. The mud itself was littered with old tyres, presumably once used as fenders but now half covered in green seaweed. The creek was full of beached pleasure craft of various sizes and in various stages of decay. There was a general air of surrender to the inevitable. Some local had attempted to capture the spirit of the place in a graffito:

WHEN GOD WAS
HANDING OUT BRAINS
I THOUGHT HE WAS HANDING
OUT MILKSHAKE SO I ASKED FOR A
THICK ONE

The boats themselves bore names that represented the hopes, loves and dreams of their owners: *Fleurette*, *Laura Louise*, *Claymore*, *Crystal Flyer*, *Cougar II*, *Pearl*, *Fairy Bulldog*. Although most of the boats were badly neglected, with bleached timbers and knocked-out windows, a few people were undertaking maintenance. On an area of hard standing I met a big, bald, scary-looking man, pale-skinned and stripped to the waist, with the build of a retired wrestler. He was wearing black boots and black shorts, and carrying a tray and a roller. He'd almost finished painting a large steel-built houseboat raised on wooden blocks. I asked him if he was a native Canvey Islander. 'Nah,' he said, staring at me with gimlet eyes. 'They've all got webbed feet.'

Leaving the boatyard behind, I cut south towards the Thames across the neck of Canvey Point. Beyond the point, looking down-river, the sky was so full of pale blue you wouldn't think there could be room for clouds. But there was. Masses of billowing early summer cumuli were sailing slowly eastward on the breeze, heading towards Rotterdam or Hook of Holland or wherever their next port of call was going to be. Extending far down the river in the direction the clouds were sailing, between the reach of the Thames called Leigh Middle and the channel off Southend called Ray Gut, stretched Chapman Sands. In the words of a local historian, the Chapman Sands once marked for London-bound mariners 'the end of a sea voyage and the beginning of a trip up river'. They also presented a serious hazard, and for a century up to 1957 here stood the Chapman Lighthouse, built, according to one of its early promoters, 'to guard the river middle'. Unlike conventional lighthouses, the Chapman Lighthouse was not a single monolith, but supported on tubular legs drilled far beneath the shifting sands and the flowing river into the bedrock beneath. The structure makes an appearance near the start of Conrad's *Heart of Darkness*: 'The sun set; the dusk fell on the stream, and lights began to appear along the shore. The Chapman lighthouse, a three-legged thing [in fact it had seven legs] erect on a mud-flat, shone strongly. Lights of ships moved in the fairway – a great stir of lights going up and going down.'

In the 1920s, the kids of Canvey used to hold swimming races out to the lighthouse, but made sure their parents and teachers never learnt what they were doing. During both world wars, convoys used to assemble off Chapman Sands, waiting for their armed escorts to join them before they embarked on their danger-ous voyages.

As I reached the southern shore of Canvey, and the main body of the Thames, I could just make out a marker buoy about half a mile out. This has taken over the warning role once played by the

Chapman Lighthouse. Faintly across the distant, alternating strips of water and sand, I could hear its bell ring. It was a mournful sound.

Judging by the housing on the south shore of Canvey, this was the desirable part of the island. Here there were rows of two-, three- or four-storey MacMansions, some executed in the Spanish style, with arcaded verandas, ballustraded balconies, pantiled roofs and walls of white stucco, surrounded by palm trees and wrought-iron railings. This, the locals would have you believe, is not saltmarsh mudflat estuary-land. This is the proper seaside, with beaches of shingle and small rocks, looking out to where great cargo ships slowly make their way upriver towards Tilbury and the London Gateway Port. Across the strips of water and the banks of Chapman Sands you can see as far as Kent, its low shore broken by oil-storage tanks and the great smoke stack of the power station on the Isle of Grain (at least it was still there when I visited Canvey; it's since been demolished). The coastal defences here are more than just a low concrete parapet on top of a bank. The sea wall here is some ten feet tall, and the shore is segmented by various species of groyne. Some of these structures are made of concrete, some are simply piles of stones. Parallel to the shore were lines of old fences, of which only a few gaunt wooden teeth remained. The sea wall bore a succession of strange inscriptions. In one place someone had stencilled a series of concentric squares; elsewhere I found this:

MO DIDDY WAS HEAR
REPPIN
CANVEY TO THE FULLEST!

The meanings of neither of these were clear to me. The one that simply said 'Dick Head' was less puzzling.

Further on I came to Concord Beach. This is a sort of lido, a series of tide-replenished paddling pools, with a low wall and railing on the sea side, and a beach of (presumably imported) sand on the landward side. The place was clearly popular with parents and kids, and there were beach towels and brightly striped chairs dotted about in among scattered flipflops and pink plastic buckets.

At the far end of the lido I found the place that sold the pink plastic buckets, together with other luminously coloured beach paraphernalia – lilos, spades, windmills, inflatable dinghies, rubber rings, toy boats. The display was crowded, gaudy and magnificent, draped all round the outside of a small concrete pavilion. The pavilion doubled as a café, so I could get a bacon roll and a cup of tea. Nearby, three middle-aged women sat smoking and drinking mugs of coffee at a green plastic table under the sea wall. Their hair was half piled up, their bodies tucked into short sundresses or shorts, their feet in white plastic sandals. I'd been taking photographs. When I got home and looked at my photos I found one of the women had been giving me the finger.

At the pavilion the woman in the Union Jack apron who ran both the shop and the café told me she'd been running the stall for twenty-eight years. But she wasn't a native Canvey Islander, she was originally from the East End. We got onto the subject of the 1953 disaster. She said that the water hadn't got in here, on the Thames side of the island where there was a sea wall, but 'over the back', somewhere on the landward side of the island, along Benfleet Creek. She drew an older man into our conversation. 'Part of it just gave in, didn't it, Brian?'

'Yeah,' he replied. 'It was just a mudbank.'

I asked him whether he remembered anything about it. Another man joined in. 'I remember it being bitterly cold.' I said he must have been very young at the time. He'd been fourteen. A friend of his had been among the dead. 'Bitterly cold,' he repeated. 'Most of the people died from exposure, not drowning. They got soaking wet, then they had to go up on the roofs.'

Now the local people were painting the sea wall with a mural commemorating the flood, and the stories associated with it. The current sea wall was built between 1975 and 1983, and at the beginning of the mural's narrative it speaks the following words:

I am your Sea Wall
I stand here 24.7.365
No matter how high the tide
Or how strong the wind
You can depend on me
And sleep safely at night
But it has not always been so . . .

One panel showed a baby in a pram cast adrift on the waters. 'That's Linda Foster,' explained the woman responsible for the panel, who happily posed for my camera, paintbrush in hand. 'In 1953 she was pushed out into the water by her parents. They both drowned, but Linda was found safe twelve hours later, still floating in her pram. That really touched me, that story.' She paused, smiled again for another photo. 'I'm proud of being part of this.'

I looked out to the Thames as a huge cargo ship made its way upriver cutting through the glassy water, releasing a powerful bow wave. I followed the ship's progress until it disappeared from sight. Some time later a series of waves hit the beach. I'd noted signs on the side of the café warning against wash. Parents were advised to look out for their children. The sea still presents a danger to Canvey and its people.

I was nearly at the end of my journey. I continued along the esplanade until I arrived at the seafront centre. There was a small hut with THAI FOOD painted on its back. Another small stall advertised 'Candy Floss Do-Nuts Popcorn'. Nearby was the low façade of Fantasy Island Amusements, its windows obscured by posters each proclaiming that this was 'Canvey's No 1 Adult Gaming Centre'. The posters advertised £500 jackpots, and asked 'Could Be Your Lucky Day?' Next to Fantasy Island, behind iron fences and barriers, was a funfair. This was 'Fantasy Island

Adventure Park', except it was all shut up and locked down. The waltzers were still and empty, the Swan Lake roundabout with its giant white plastic swans was stationary and silent, the slides were packed-up close to the green, pink and yellow dolphins, the shooting galleries in mothballs. The nearby 'Visitor Information Centre and First Aid Post' was also closed.

The place that *was* busy was the Monico, an elegant white-painted Art Deco pub draped with England flags. It was heaving. There was a sign outside:

> B.O.B.B.
> Behave Or Be Banned
> MEMBER
> We Exclude Troublemakers From
> These Premises

Another sign read:

> Please NO SWEARING
> AS IT OFFENDS

Someone had painted over half of the sign.

I was relieved to be ignored by the clientele when I entered the Monico. A row of older men in summer casuals – shorts and sleeveless shirts or t-shirts – lined the bar, each spread out on his own stool. There were more England flags, and behind the bar a big screen stood ready for the World Cup. I'd run out of time, had to be back in London by the early evening. So I asked the barman if he could phone for a minicab to take me back to the station at South Benfleet. He was happy to oblige.

I asked the minicab driver if he could take me past the oil-storage depot in the southwest of the island. As he stopped to let me take some photographs, he told me that the giant tanks had once been used to store the corpses of cattle killed by BSE. Now much of west Canvey is a nature reserve, he said, home to various endangered bird species. There was not much else in this part of the island apart from the cemetery, he said.

The year after my visit to Canvey, I learnt that the island was once again to become a dumping ground for unwanted corpses, this time human rather than bovine. In 2015, the engineers digging a tunnel for Crossrail under the site of the old Bedlam Hospital at Liverpool Street Station found they had 3,300 skeletons on their hands. The London cemeteries were full, so the old-established East London firm of T. Cribb and Sons, Undertakers, was tasked with finding an affordable plot. This they did, and the remains of the deranged dead of Bedlam – some of them victims of plague as well as madness – are now interred in Willow Cemetery on Canvey Island. It was a strange twist in the story of Canvey's settlement by outsiders – from Danes and Dutchmen to madmen and mad cows.

In the EU referendum of June 2016, 72.7 per cent of those who voted in Castle Point, the local government area that includes Canvey Island, voted 'Leave'. Only two other areas in England had higher percentages voting for Brexit. The previous year, Castle Point had a net inflow of eighty-one new international migrants. In contrast Lambeth, which recorded the highest 'Remain' vote of 78 per cent, the figure for the same period was 4,598. So Canvey Island could hardly claim it was being 'swamped', to borrow the kind of language used by those exercised by immigration. But perhaps it was fear of being overwhelmed by the unknown, the other, the alien, that prompted the strong 'Leave' vote. The people of Canvey had lived for long enough with the threat, and the reality, of being overwhelmed by the sea, the ultimate in otherness. After all, the islanders were human beings, not web-footed amphibians as outsiders averred. Then they had had to accommodate the other in the form of thousands of non-islanders from the East End, who were in turn fleeing the 'ethnic minorities' who had moved into their areas. Once they were on Canvey, the much-expanded population not only raised many flags of St George, but also the metaphorical drawbridge that connected them to the rest of the world. If Britain was no longer an island, it seems they were determined that Canvey should remain one. In their own eyes at least, they are the last of England.

~

Except it isn't quite that straightforward. Three years after visiting Canvey and some time after writing the bulk of this chapter I learnt that a number of strictly observant Haredi Jews from the large community in Stamford Hill, north London, squeezed out by ballooning property prices in the capital, have started to buy houses on Canvey Island. By all reports, the locals have welcomed them warmly.

It was Chris Fenwick, aka 'Mr Canvey', local entrepreneur and long-term manager of Dr Feelgood, who eased the acceptance of the new arrivals. He'd first encountered two of them – 'the two Abrahams' – outside the Lobster Smack. They were on a scouting expedition. 'It was a hot day and there they were in the full clobber, looking out at the estuary with the ships coming by,' he told the *Guardian*. 'So I walked up to them and said: "Shalom, gentlemen, what brings you to Canvey?"' He went on to show them the cover of Dr Feelgood's album *Down by the Jetty*, which features a photograph of the band standing in the exact same spot where he'd first seen the scouts.

'It's all too spooky for words, boys,' Fenwick had said.

And they'd replied, 'No, no, no. It's divine intervention.'

Once some of the Haredi families had settled on Canvey, Fenwick hosted a meal at his hotel for both Haredis and Canveyites. The new arrivals were somewhat bemused when someone belted out a rendition of Dr Feelgood's hit, 'Milk and Alcohol'.

But on Canvey you have to expect the unexpected.

Through the Red Steel Door

Eel Pie Island

Miss Morleena Kenwigs had received an invitation to repair next day, per steamer from Westminster Bridge, unto the Eel-pie Island at Twickenham . . .

– Charles Dickens, *Nicholas Nickleby* (1838–9)

Upriver from its estuary, its waters thickened by soil washed off the fields of southern England, the Thames has spawned a number of small islands. Known locally as aits, or eyots, these teardrop-shaped agglomerations of silt are elongated in the direction of the river's flow. As reeds and sedges, then alders and willows, take hold on the newly formed land, the roots trap more silt, and the ait grows. Aits may only be temporary, and when slowly washed away the sediments from which they were once formed may coalesce into another ait downstream. The islands in the Thames Estuary, such as Canvey, may be partly formed from smaller islands that once existed much further upriver.

The first of the river islands in the Thames, travelling upstream from London, is Chiswick Eyot, accessible by foot at low tide but largely submerged when the tide is high. In September 2010 a pensioner called Nick claimed to have been living alone on the island for the previous six months, sleeping in a bed he'd built in the branches of a willow tree, above the high-water mark. 'Home, sweet home,' he told the BBC. 'It's better than any bed I've slept in.' Dressed in shorts, wellington boots and a black ex-army beret, he would wade with his bicycle to the mainland at low tide to collect supplies, but warned that you have to be careful of the speed of the incoming flow. 'I've achieved freedom,' he replied when asked why he'd chosen this life. 'Things are now simple, and uncomplicated. You don't have to go and knock on any doors. Most people if they had to choose between being here, on an island, or being in a bank, or a government office, would say, "Ooh, give me the island."' Hounslow Council, which owns Chiswick Eyot, insist that the island is uninhabited.

Further upstream there is another uninhabited island, originally known as Strand Ayt but now called Oliver's Island, from the rumour that Oliver Cromwell had once sought refuge here, reaching his headquarters at the Bull's Head in Strand-on-the-Green via a secret tunnel. On the other side of Kew Bridge is the larger Brentford Ait, once home to a 'House of Entertainment', the scene of goings-on that upset the locals. In 1811, a Mr Robert Hunter of Kew Green complained that the island 'has long been a harbour for men and women of the worst description, where riotous and indecent scenes were often exhibited during the summer months on Sundays'. Brentford Ait is now bereft of men or women of any description. In the 1920s it was planted with trees to shield visitors to Kew Gardens from the view of Brentford gas works.

Above Brentford Ait there are other temporary patches of dryish land in the midst of the Thames's flow. Isleworth Ait, Corporation Island and Glover's Island provide homes to willows, herons, ducks and cormorants. Then, near the top of the tidal Thames, above Kew and Richmond, a little short of Teddington Lock, squats what was first recorded in the fifteenth century as Goose Eyot. The river here may still have a tang of salt from the North Sea in its waters, so this might just count as the furthest inland of all of England's coastal islands.

Goose Eyot – once comprising three, then two, separate islands – was later known as the Parish Ayte, then Twickenham

Ait. By the early nineteenth century it was celebrated as Eel Pie Island, in honour of the delicacy on offer at the island's White Cross Inn, which had become a magnet for boating parties up for the day from London. (The proprietrix of the inn claimed that Henry VIII was wont to stop by for eel pie on his way by royal barge to Hampton Court, but it seems this story was no more than a canny marketing ploy.) By the late 1830s, when Dickens published *Nicholas Nickleby*, trippers were arriving at Eel Pie Island, not by rowing boat, but by steamer:

> It had come to pass, that afternoon, that Miss Morleena Kenwigs had received an invitation to repair next day, per steamer from Westminster Bridge, unto the Eel-pie Island at Twickenham: there to make merry upon a cold collation, bottled beer, shrub and shrimps, and to dance in the open air to the music of a locomotive band, conveyed thither for the purpose: the steamer being specially engaged by a dancing-master of extensive connection for the accommodation of his numerous pupils . . .

Dickens does not record how long it took to travel by steamer from Westminster to Eel Pie Island. It took me more than two hours to travel from Hornsey, my home in north London, to Twickenham, the nearest station to Eel Pie. It was one of those extended Transport for London epics, involving bus, tube and two overground trains.

Twickenham Station and the roads around it have all been rebuilt since 1964, when Long John Baldry found Rod Stewart playing 'Smokestack Lightnin'' on his harmonica on the platform while waiting for a train. Impressed, Baldry asked Stewart to join his band. Baldry was one of the stalwarts of the Eel Pie Hotel, which in the 1950s had been an important trad jazz venue, featuring such acts as Kenny Ball, George Melly and Acker Bilk, and had since – as the 'Eelpiland' club – become the home of British rhythm and blues. The Yardbirds, the Rolling Stones, the Who and John Mayall's Bluesbreakers all played here, alongside Baldry's Hoochie Coochie Men. 'Any band that was worth its salt had to play there,' recalls Paul Jones of Manfred Mann. 'Till you ticked off that one on your itinerary, you hadn't really arrived.' For young Londoners, going to Eelpiland, with its sprung wooden dance floor, was something of a rite of passage. Many years later the American actor Anjelica Huston, who did some of her growing-up in London, remembered the atmosphere with some ambivalence: 'The room would just be throbbing – hot, humid, full of cigarette smoke. People didn't take a lot of baths in those days in London. The music'd be blaring. Those who weren't dancing were snogging. It was a kind of ritual thing.' By 1967 the hotel had fallen into disrepair, and was closed down by the authorities. Squatters moved in, tearing out bits of wood and burning it to keep themselves warm through the winter. The building itself burnt down in 1971.

Walking down London Road from Twickenham Station I heard no echo of a locomotive band, no faint wail of the blues. Instead, there was just the noise of the traffic grinding past a parade of the usual names: Burger King, Costa, Waitrose, WHSmith, Specsavers, Betfred the Bonus King. It could have been any high street in England. Towards the river, though, the houses became older, more Dickensian. One boasted a large black lantern hanging off its exterior, proclaiming in stencilled white letters

PAINE

FUNERALS

I cut down Water Lane, away from the traffic towards the river. It was a different world. Soft January light glinted off the surface of the Thames, flickering between brown and gold. The air was filled with the cries and fretful flappings of gulls. Canada geese stood waiting to be fed on the hard standing by the water's edge. A child threw bits of bread. Not far away across the channel, Eel Pie Island waited, packed tight with boats and trees and houses.

There's only one way across for pedestrians: a high arched metal footbridge. It clanged and vibrated as I walked over. On the far side I was greeted by a large sign that said:

PRIVATE ISLAND
NO THOROUGHFARE
OR ACCESS TO THE RIVER
NO CYCLING

As I stepped onto dry land, there was a noticeboard carrying an advertisement for 'art lessons', while another announced that a 'Mature single F' wished to rent a studio/flat on the island. Then there was a schematic map of Eel Pie, marking the small, individual houses, many built of wood and bearing such names as Shamrock, Pie Crust, Copper Beech, Desdemona, Wild Thyme, Wyndfall, Blinkwater. There was a certain quaintness in the air. Close to the map, a house called the Nook boasted a nautical porthole in its door. It stood at a corner by a path blocked by a wrought-iron gate, warning 'strictly private' and 'residents only'.

In the opposite direction a surfaced path curved past fences and gates, its borders planted with shrubs. A succession of signs announced 'CCTV', 'No entry', 'Beware of the dog', 'Keep out'. The Lion Boathouse, with its red door and green corrugated iron walls, tried to soften its prohibitions with humour, bearing an old metal notice that read 'Any person omitting to shut and fasten this gate after using, is liable to a penalty of forty shillings.' There was no gate.

I passed a few people walking along the path. Most seemed to be in a hurry, gazing straight ahead, perhaps recognising a stranger and hoping to ignore him. I caught the eye of one man who was walking slowly enough for me to engage him in conversation. He was dressed in black, must have been in his later sixties. He told me he'd lived on the island for twenty years. Sometimes, he said, it was an advantage living on an island, sometimes not. It could be a long walk to anywhere. He pointed out a rather dull block of 'town houses', and told me they'd been built on the site of the Eel Pie Hotel. His sister used to go to gigs there when she was young – before his time, he assured me. She'd dress up in very tight jeans, then sit in the bath to make them shrink even more tightly round her figure. She'd had kidney problems ever since as a consequence.

I asked whether there was some kind of social centre on the island, since the hotel had burnt down. Just Richmond Yacht Club, he said. Many of the residents belonged to that. He didn't.

The path I'd been following soon came to an end at a large building with a red steel door. The building looked like a warehouse. The door was shut. 'No Public Access,' it said. 'Private. Working Boatyard.' My visit to Eel Pie had barely lasted fifteen minutes. There was nowhere else on the island I could walk. Most of the land was private and prohibited. For a place supposedly full of artists, and formerly home (in the late 1960s) to the UK's largest hippie commune, Eel Pie Island seemed to have retreated into fenced-off smugdom.

Disappointed, I turned on my heel and walked back across the bridge to the mainland. Four boats from the rowing club powered upstream towards a sky burnished with grey and silver. Back on the mainland the child was still feeding the birds at the water's edge. The geese had been joined by pigeons and a pair of mute swans. Three elderly Polish men stood by a bench, huddled in winter coats, cracking open cans of high-strength lager.

I followed the path along the left bank, then over a steeply stepped and ornately balustraded footbridge into the grounds of York House, an elegant brick-built seventeenth-century mansion. My intention was to walk to the station at Richmond, crossing the river at Hammerton's Ferry. A boat trip was needed, I felt, to make this a proper island day, even though the ferry did not connect with the island. My plans were initially thwarted, as I found there was no way back out of the grounds of York House other than to retrace my steps. The grounds were a maze of paths, ponds and yew hedges. At the end of one cul-de-sac there was an extravagant display of statuary. This display, I later learnt, is popularly known as *The Naked Ladies*. It consists of several over-lifesize Oceanids carved from Carrara marble. They are frozen in various postures of abandonment, adoration or distress, draped over a chunky rockery running with cataracts and verdure. The statues had originally been commissioned by the fraudster Whitaker Wright, who in 1904, just after being sent down for seven years at the Royal Courts of Justice, committed suicide by ingesting a cyanide capsule. The statues were then purchased by Sir Ratanji Dadabhoy Tata (of Tata Steel fame), who installed them in the grounds of York House, which he had purchased from the Duc d'Orléans in 1906. During the Second World War the authorities covered the statues in grey sludge, for fear that moonlight reflecting off their white marble bodies would provide German bombers with an aid to navigation.

Saying farewell to the sea nymphs, I hastened away to meet up with the ferryman. Retracing my steps I reconnected with the narrow road called Riverside. This road, several signs warned me, was 'liable to sudden flooding'. As if to prove the point, a canoe half filled with water was drawn up on the shore. High tide, another sign advised me, was in about an hour's time. Dusk would not be that far behind, and dusk was when the ferry would no longer run. I picked up pace.

In 1908 a licensed local waterman called Walter Hammerton began to run a ferry service for foot passengers across the Thames between Marble Hill, just downriver from Eel Pie Island, and Ham on the other bank. This was in competition with a more established ferry half a mile upriver, which had been started in the seventeenth century by the local lord of the manor. His descendants had become Earls of Dysart, and in 1913, in response to Hammerton's entrepreneurial impertinence, a case was brought against the upstart waterman by William John Manners Tollemache, the Ninth Earl, whose principal seat was Ham House. Lord Dysart – whose wife had left him some years earlier owing to his eccentric and cantankerous nature – lost the case, but won on appeal the following year, and Hammerton was obliged to cease operation. However, a public subscription was launched to support his case and in 1915 the House of Lords found in favour of Hammerton. His ferry has run ever since.

Near to the quay where Hammerton's Ferry berths on the left bank there's a photo of a dapper-looking Walter Hammerton standing at the stern of his ferry. He wears a suit, tie and trilby, sports a neat little goatee, and holds a sculling oar in his right hand. On the ornately carved back-rest of the passenger seat in front of him fancy lettering describes how he 'won this ferry for the public'. Since Hammerton retired in 1947, the ferry has had a succession of owners.

When I walked out along the wooden pontoon to board, two passengers were disembarking from the wide, flat-bottomed motor launch that has replaced Hammerton's rowing boat. A young man was at the wheel, wearing a multi-coloured knitted jacket with the hood pulled up against the chill afternoon. I was the only passenger. I handed over my £1 fare, and asked him whether it was the family business. 'Yes,' he said. 'It's me dad's business. I'm more of a water gypsy meself.' I told him about my somewhat abortive visit to Eel Pie Island, brought to a halt by the red steel door saying 'No public access'. 'Ah,' he said, 'you want to go through that door, it won't be locked, no one'll mind. Beyond the boatyard is the interesting bit. Artists' studios. Hippies. You want to go through that door.' By this time we'd reached the far bank – the river is not wide here. I said I'd better go back before it got dark, and pressed another pound on him for the return fare. 'Don't worry about that,' he said. I insisted, saying he had a living to make.

I hurried back along the shore towards the bridge to the island. To the southwest piles of cumuli were towering up over the river as if some great drama was about to unfold. Gulls wheeled about, alert and anxious. The bridge clanged under my feet as I strode quickly across, passing the sign saying 'Private Island No Thoroughfare'. I followed the curving surfaced path between the fences and shrubs, not pausing this time to note the names of the houses, or their signs of welcome. In no time I was in front of the red steel door. It was closed, but not completely shut. I pushed. It yielded. I half expected to hear a shout of 'Oi! You!' but there was only silence. Inside the cavernous, high-roofed building two or three houseboats had been drawn up out of the water for repair. Ladders, hosepipes, steel struts and piles of timber were scattered about.

This being a Saturday, no one was working in the boatyard. I walked out through the open side of the building opposite the red steel door. There was an alley lined with painted wooden shacks, pot plants and dangling lifebelts. On one door was written the name 'Eel Pie Bombers'. Underneath this was painted a two-legged but otherwise mermaid-like redhead astride a 500 lb bomb, a bomb in the guise of a bloody-toothed whale. The redhead was naked bar her red-and-white striped socks and a star-spangled bikini top. She looked like she might have once adorned a B-17 Flying Fortress, ready to rain down hell on Hamburg or the Ruhr. I wondered why those young American airmen so many years ago had wished to have their bombing raids blessed by a sex goddess. The British had, after all, covered up their sea nymphs at York House with grey sludge, for fear their naked beauty would glint in the moonlight and lure the Luftwaffe like sirens.

I peered through some windows. One shack housed a pottery, while others were stuffed with clutter that the owners no doubt thought might one day be turned into meaningful works of art. There was nobody about, just a couple of manikins staring blankly over my shoulder, one clutching a can of Carlsberg with a rose stuck in it, another wearing nothing but dark glasses, pigtails and a horned helmet. Tables and chairs were scattered here and there outside the shacks. Some of the tables bore bowls of oranges, others unlit candles and empty champagne bottles, as if there'd been a party some time in the past. It all looked like everything that had ever happened here had happened a long time ago.

I wandered down towards the water's edge, where many craft were moored, several abreast, awaiting repair. No one was working on them, no one was around at all. Then a black cat appeared at the entrance of a narrow gangway leading over the water towards the boats. The cat disappeared into the shrubs, and was replaced by a solitary woman carrying plastic bags full of shopping. She was dressed in black: black jumper, black jeans, black boots. She had long dark hair, a soft voice, a tentative smile.

I explained myself, and my presence on Eel Pie. I asked whether, as the island was in tidal waters, it counted as coastal. 'Of course,' she said. Was there any salt to be tasted in the water? 'That I don't know about,' she laughed. 'I've never tasted it. A friend fell in a little while ago and was rather ill afterwards.' I asked her about the tides. She told me the road, and the cars parked on it, on the mainland by the strand were often flooded. She and the other river-dwellers were safe on board their boats, they just floated. She pointed to a marker post. 'When it gets to the top of that, we start to get worried.' She explained that these weren't technically residential moorings. She was just moored here because her engine was being repaired. She'd been at Eel Pie two months so far. They could get electricity and water and sewerage, and after the visit of a coal barge in November she'd had enough fuel to keep her warm through the winter. She knew all about the island's musical history. Her father had played with Long John Baldry.

She pointed out a large barge, told me the owner sailed it down the Thames to London to watch the New Year fireworks. He'd even been in it to Holland and back. 'So you're not insular here, even though you're on an island?' I asked.

'Oh no, we're not insular. We're open to the Continent.'

The light was getting low, the tide approaching the full, as I made my way back onto the mainland. The strand on the shore was now underwater. No one was left to feed the birds. Gulls filled the air, forever unsettled, while behind them, to the west, dark castles of cloud swelled up into a silver-streaked sky, an unlikely, apocalyptic backdrop to London's western suburbs. As I walked inland back towards the station, I sensed behind me the great river – indifferent to the works of man, obeying only the law of gravity – surging eastward between its banks, around its aits, irresistibly, silently, to the sea.

You'll Have a Blast

The Isle of Sheppey

Full of *crazy* people.

– Landlady of Old House at Home pub, Queenborough

I met Baz the fisherman on the quay by the creek called the Creek at Queenborough. I'd got off the little local train one stop before its terminus at Sheerness. My intention was to make the last mile or so of the journey on foot, along the Medway shore of the Isle of Sheppey.

Baz was untangling nets, hauling yards and yards of blue-green plastic line from the back of a pick-up truck. I'd passed by a boat sitting on the mud of the creek with bunches of flowers tucked all along the taffrail, around its cockpit and cabin. I asked Baz about the flowers. 'The old dad died,' he told me. 'He owned the boat. Now his sons, on birthdays and other special days, decorate it with flowers.'

There weren't many working fishermen left on the Isle of Sheppey, he said. 'In the eighties there were fourteen boats running out of here commercially. Now we've only got two or three.' He fished mostly for skate, and worked the local beds of whelks, mussels and oysters. There were also cruises for sea anglers and seal-watchers in his boat *Stella Spei* ('star of hope', he explained) and trips out to the Red Sands Towers. The Towers were Second World War anti-aircraft forts positioned in the middle of the mouth of the Thames Estuary. 'The men were posted out there for six weeks at a time,' he said. 'They had some pretty big guns.' He told me many German bombers just dropped their loads over the sea and turned for home rather than run the gauntlet of fire they'd face as they flew up the Estuary towards London. So there were a lot of unexploded bombs scattered across the seabed.

'Also you've got the *Richard Montgomery* wreck just off Sheerness,' he added as an afterthought. 'American munitions ship.'

'The one where you keep your fingers crossed?' I asked. I'd heard about the *Richard Montgomery* before my journey to Sheppey, and felt its faint threat hang over me as I approached Sheerness. 'You don't go too near that do you?'

~

In August 1944 the SS *Richard Montgomery* – a US Liberty ship named after an American patriot killed fighting the British in 1775 – took on a cargo of munitions at Hog Island, Philadelphia, before sailing down the Delaware River and across the Atlantic. When it arrived at the mouth of the Thames Estuary, the harbour master at Southend, in charge of all shipping movements in the Estuary, ordered it to take up a berth in the Great Nore Anchorage while it waited for a convoy to assemble. Then it was to sail across the Channel to Cherbourg, liberated by the Allies just a month earlier.

On 20 August, while its captain slept, the *Richard Montgomery* dragged its anchor. Several ships nearby saw it shifting, and sounded their sirens in warning. To no avail. The *Richard Montgomery* ran aground on a sandbank at the entrance to the Medway channel, just a mile and a half off Sheerness. When the tide fell, the ship broke its back. Attempts to salvage the cargo were abandoned a month later when the hull snapped in two, leaving some 1,400 tons of high explosive still on board. The munitions included – and still include – 286 'blockbuster' bombs, each weighing 2,000 lb, and so named because each one was – is – capable of destroying a whole block of buildings. The wreck also contains several thousand 1,000 lb bombs and 2,500 cluster bombs with their fuses in place.

The ship's three masts can still be seen sticking above the water at all states of the tide. The whole area round the wreck is an exclusion zone.

The authorities insist that there is little risk of an explosion. Others say it is a real possibility. In 2004 *New Scientist* concluded that if the ship went up, a column of debris would be thrown nearly two miles into the air, a tsunami would power up the Thames towards London, Sheerness itself would be devastated, and, if hit, the tanks of liquid gas on the Isle of Grain, across the mouth of the Medway, would unleash a further inferno.

Experts say that the explosives should remain inert unless exposed to sudden shock, friction or heat. But the annual surveys by the Maritime and Coastguard Agency suggest the wreck is slowly disintegrating, so a sudden collapse leading to a mighty detonation cannot be ruled out. There have already been a number of near misses with vessels in poor visibility. 'If this boat had gone down outside the Houses of Parliament,' complained one local councillor, 'something would have been done long ago. How far down the river do you have to go before a dangerous wreck becomes acceptable?'

At the edge of the town of Sheerness there's a mural that greets visitors. It depicts a sour-faced mermaid lying on a beach with the masts of the *Richard Montgomery* emerging from the sea behind her. The mermaid has one hand on a plunger marked 'TNT'. The wire leads across the beach and into the sea, towards the wreck. The mural declares

Welcome to
SHEERNESS
You'll have a blast

~

'You don't go too near *that* wreck do you?' I asked Baz.

'Well the funny thing is,' Baz said, 'when we're dredging for mussels, any bombs that we pull up, we'll wait till we've got half a dozen on board and then we drop 'em round the perimeter of the wreck, so we don't keep picking the same bombs up again and again.'

I asked whether it wasn't a bit nerve-wracking hauling up all this dangerous material.

'A little bit,' he conceded. 'The big ones can be a worry. Years ago they used to just drop 'em back down, but what we do now is call up the MoD. They'll come out. They pay compensation for your down time. Years ago it weren't like that.'

Sometimes there were naval mines, he said, 'the big ones with spikes coming out of 'em'. Sometimes he pulled up pairs of cannonballs linked together with a chain, of the sort used in Nelson's era for snapping the masts of enemy ships, or the heads off enemy sailors. Nelson had trained at Sheerness, he told me, and is said to have shared a house in Queenborough with his mistress, Emma Hamilton. After Nelson's death at Trafalgar, his body, preserved in a barrel of brandy, was supposedly landed at Sheerness. The naval dockyard there had been established by Samuel Pepys in the 1660s, and had only closed in 1960. The area round about the docks was called Bluetown, after the colour of the Royal Navy's paint the workers had appropriated to decorate the huts they lived in. 'Later,' Baz said, a little sheepishly, 'Bluetown was full of, well, you know, pubs and brothels.'

Being on the edge of England, the Isle of Sheppey has found itself in a number of England's wars. Edward III built a castle at Queenborough to guard his supply ships making their way to France during the Hundred Years War via the Swale, the sheltered channel that runs west and south of the island, separating it from the mainland. The town is named after his queen, Philippa of Hainault.

Baz remembered that the Dutch had occupied the island – or at least Queenborough and Sheerness. This was in 1667, during the

Second Anglo-Dutch War, when the Dutch sailed up the Medway and destroyed the Royal Navy at anchor off Chatham, one of the greatest humiliations in English military history. The nation had, one contemporary mourned, 'lost its honour at sea for ever'. Although the Dutch only held the Isle of Sheppey for a few days, more than one local told me that Dutch rule had lasted until the 1960s. In fact, in 1967 there *was* an official ceremony at which the Dutch formally handed back Queenborough to the UK.

By this time, or some time before, Queenborough had become a bit of a backwater. In the early eighteenth century Daniel Defoe described it as 'a miserable and dirty fishing town'. In the nineteenth century, prison hulks were moored offshore, and the corpses of dead convicts were dumped on a nearby salt marsh. The place became known as Deadman's Island. Rising sea levels and coastal erosion are leaving unmarked coffins and bits of bone sticking out of the mud, visible at low tide. Access to the island is prohibited.

~

I asked Baz where I might get a bite of lunch. He said there were a couple of pubs along the High Street, one called the Dutchman and the other Old House at Home – without a 'the'. It was not to be confused with the Old House at Home (with the 'the') in Sheerness. I opted for Old House at Home, looking out across the mouth of the Swale and the Medway beyond.

There were perhaps eight men sitting under a sunny window in the confined interior. Soon after my arrival they downed their pints and left. I asked the landlady if I'd done something wrong. 'It's a Wednesday ritual,' she reassured me. 'They come in on their boats, pop in for a pint, then go and who knows what.' She was a handsome, strongly built woman, confident and friendly, with tied-back dark hair. I ordered a pint, and steak and kidney pudding. There was an option of gravy or ketchup. I chose gravy. She told me I'd made the right choice.

There was only one other customer in the pub now, sitting on a stool across the corner of the bar where I perched. He had short grey hair, a grizzled grey beard and steel-rimmed spectacles. He was a musician, he told me when I asked, Sheppey born and bred. He played the local pubs and clubs, either solo or with a band. 'Middle of the road stuff,' he said. His manner was taciturn, his air melancholy, his eyes moist. I thought he must be in his sixties, but he was actually ten years younger.

I asked him whether that was the Isle of Grain across the water. 'Yes,' he said, and returned to his silence. Wasn't there a big power-station chimney there, I asked, remembering that I'd looked across at it from Canvey Island three years before. 'They blew it up,' he said.

Later I found a video online of the demolition, at 11 a.m. on 7 September 2016. One moment the chimney is there, thick and strong and over 800 feet tall. Then a thin ring of smoke erupts around the circumference of the chimney about half way up, and the whole structure falls down neatly on itself. Where the chimney stood in the sky is marked by a chimney-shaped column of dust, pouring upwards out of the rapidly falling mouth. Then the column of dust itself collapses and spreads in a slow rolling wave across the ground.

The Isle of Grain is no longer an island, but forms the marshy eastern point of the Hoo Peninsula on the other side of the mouth of the Medway, facing the Isle of Sheppey. I asked the musician if there was a ferry linking the two. 'No,' he said, 'you have to go the long way round.' Then he volunteered the fact that the Isle of Sheppey *had* been connected by passenger ferry, not to the Isle of Grain, but to Southend, on the other side of the Thames Estuary. 'They tried it a couple of times over the last twenty years, but I don't think it was financially viable. Bit of a shame. It's quicker than driving. It had a little bar on it. People used to shoot over to Southend for the day.'

I told them about my English islands project, said I'd visited Canvey Island, next to Southend.

'Stick around,' said the landlady. 'You ain't seen nothing yet.' And she fell about with laughter. The musician joined in.

'Compared to Sheppey?' I inquired.

'You'll write a *big* book,' said the musician.

'Well, Sheppey's a big island,' I said.

'Full of *crazy* people,' said the landlady, and began to cackle and hoot.

'You gotta stick around,' said the musician.

'When does it all kick off?' I asked.

'Any minute,' said the musician. The landlady was still laughing helplessly.

'So I've come to the right place?'

'Oh yeah,' said the landlady, 'most definitely.'

In a while the hilarity subsided back into silence. The landlady, having established again that I was a gravy man, returned to the kitchen.

'So what do most people do these days?' I asked, to get the conversation moving again. 'Work in the docks, run caravan parks?'

'Yup,' said the musician.

I said I'd read there was a factory in Sheerness that makes garden gnomes.

'That's right,' said the musician. 'Whelans, that is. In Bluetown.' Whelans not only makes gnomes, but a whole range of concrete garden ornaments, from nymphs and Buddhas to owls and dragons. 'You go that way, keep on walking,' the musician continued, 'and you'll come to Bluetown, and that's where the dockyard is, the gnome factory . . .'

We fell into another silence, until I asked, 'What was it like, being a kid here?'

'Um,' the musician pondered. 'It was all right, yeah. It was a lot less busy than it is now.'

'So what's it busy with?' I asked.

'Oh, just people coming down from London, moving in.'

'So there's more people living here permanently?'

'Yes. I was born here in the Sixties. Now there's too many houses, too many people. Ridiculous. It's not like it's a drive-through place. It's an island, so the whole thing just gets clogged up.' The musician paused. 'I think, anyway.'

I concentrated for a while on my steak and kidney pudding. 'So you went to school here?' I asked. 'In Sheerness?'

'Cheyney,' a new voice said. We'd been joined at the other end of the bar by a young builder and his dad, for whom he worked. Both were wiry and tanned. It was the son who'd spoken.

'How d'you know that?' asked the musician.

'Cos I went there too, mate. Me and me dad. Cheyney Middle School.'

'Blimey,' the musician might have been impressed. 'There you go then.'

The young builder, hearing of my project, started to list places of interest on the island. 'Warden Bay beach, quite a lot of people come from off the island to do fossiling there. Another place is Harty Ferry, down the Swale. Sometimes I go there with my bins and you can see people in Faversham drinking their pints, it's so close.'

'At low tide you can pretty much walk across,' offered the musician.

'The tide comes in incredibly fast there,' said the young builder. 'I was only gazing for ten minutes and it was filling me boots up.'

I was trying to find Harty on my map.

'It's past all the prisons,' said the young builder.

'There's prisons on the island?' I asked.

'Three of them,' said the young builder.

'This is like Alcatraz, mate,' said the musician.

The young builder's dad broke his silence. 'You come on 'ere, you don't never come off.' I thought of the convicts of long ago who only escaped the prison hulks when their bodies were dumped on Deadman's Island.

Today there are some 2,800 inmates in the island's three prisons, a significant proportion of Sheppey's total population of around 38,000. Among those who have served time here is the perjurer and former Tory cabinet minister Jonathan Aitken. Another inmate was Michael Bettaney, the MI5 agent who was sentenced in 1984 to twenty-three years for passing secrets to the Soviets.

The young builder was more interested in the parts of the island still left to nature. 'Pretty much half of it is natural. Bird sanctuaries, hides, loads of marsh harriers. Pretty wild. Especially when the fog comes in and sits on the marshes . . .'

Then he too drifted off into silence, as if the fog had settled on his train of thought. I imagined the tide filling his boots, recalled Pip's first terrifying encounter with Magwitch in a north Kent graveyard not far from where I sat – Magwitch, the convict who has just escaped the hulks, but not his shackles, 'A man who had been soaked in water, and smothered in mud, and lamed by stones, and cut by flints, and stung by nettles, and torn by briars; who limped and shivered, and glared and growled; and whose teeth chattered in his head . . .'

Then in a while the fog of melancholy drifted away. The talk revived and turned to the Dutchman, the other Queenborough pub, just along the road, close to the Creek. The musician remembered how the landlord used to organise all sorts of entertainments: 'Like walking the plank. D'ye remember that? Captain Cutlass. One guy jumped off the sea wall one summer inside a washing machine.'

'How was he afterwards?' I asked.

'A bit shaken up,' answered the young builder.

'Spun out,' said the musician. 'Dried himself off nicely.' There were a few cackles. 'No, it's the truth. They'd jump off in all sorts of stuff. Got quite dangerous in the end. I think that's why they stopped it.'

'So it was a competition to do the craziest thing?' I asked.

'Yeah,' confirmed the company to a man.

'So is it being on an island that makes people crazy?' I asked.

'Living on *this* island,' said the landlady.

The musician agreed. 'It does make you crazy.'

'But there's lots of ways you can get off the island?' I asked.

'Oh yeah,' said the musician. 'You can get away from the insanity of it.' Then the door opened and two men entered. 'Oh-oh. Here comes some now. Prime example. Two crazy people.'

'I resemble that remark,' said one of the newcomers, a large man in overalls with a West Country accent. 'You've got to be a bit insane to survive this place.'

I was thinking of Captain Cutlass, of pirates and the past. I asked whether there had been lots of smuggling around the island over the centuries.

'No,' the other newcomer said, a smaller man, with what sounded like a Dutch accent. 'This was a Dutch island until the 1970s.'

I asked him whether he was Dutch. German, he said. I asked what had brought him to Sheppey. Twenty years ago, he said, he'd been working in Hong Kong, until they threw him out for not having a work permit. The cheapest flight back to Europe was to Heathrow. He'd lived for a while on a friend of a friend's decommissioned motor torpedo boat. He'd put a new deck on it. Then he'd got himself a sailing boat. He'd wanted to moor it at Southend. 'But in Southend,' he told me, 'you've got no deep-water moorings, except in Ray Gut round the corner. So I turned up here. After a big storm.'

'And the rest is history,' said the musician. He called the German 'Fritz'. I don't know whether that was his real name.

'Sheppey attracts flotsam and jetsam from all round the world,' someone said.

There was general laughter.

'Not very flattering,' said the German. 'But yes.' He turned to me. 'You know what they call them?'

'Who?' I asked.

'The people of the Isle of Sheppey. Swampies. That's what they call them.'

'We don't mind,' someone said.

'Some of us do,' said someone else.

Anxious to change the subject, I asked whether the island had any haunted places.

'Loads of ghosts,' said the landlady.

'There's one in here,' someone said. 'Some previous landlord.'

'There's actually two,' said the landlady. She turned to the musician. 'Your dad. He walks the place, his dad. Often see him wandering about.'

'So he was the landlord here, was he?' I asked.

'Yeah,' said the musician.

'He's still about,' said the landlady, 'keeping his eye on things. If I'm down here at one or two o'clock in the morning, on me own, I see him floating about.'

'I couldn't handle it,' said the musician.

A silence fell, until someone said, 'Friendly ghost.'

'I don't know about that,' said the musician.

He took a sup of his beer. 'That's him up there.' He pointed to a photograph above the bar. It showed a man with a neatly trimmed dark beard part-covering a fresh, plump face. He was wearing a white captain's hat. 'Coxswain of the Sheerness lifeboat, my dad. Charlie Bowry, that was his name. He was involved in lots of rescues. Radio Caroline, that was one of 'em.'

That was back in 1980. The ship in question was the *Mi Amigo*, which had begun life in Germany in 1921 as a three-masted schooner called *Margarethe*. After the Second World War it became a pirate radio ship, eventually for Radio Caroline, anchored at various places off the Essex coast. On the night of 19 March 1980, in a northeasterly Force Ten, the ship's anchor-chain broke. The *Mi Amigo* began to drift, and went on drifting for miles. All this time, Radio Caroline continued to broadcast, even as the Sheerness lifeboat came alongside in high seas to evacuate the crew. All four men aboard (two of them DJs) were rescued, together with a canary called Wilson, named after the former prime minister. Just before midnight, Radio Caroline stopped broadcasting. DJ Stevie Gordon had the last word: 'We're sorry to tell you that due to the severe weather conditions and the fact that we are shipping quite a lot of water, we are closing down, and the crew are at this stage leaving the ship. Obviously, we hope to be back with you as soon as possible, but just for the moment . . . from all of us, goodbye and God bless.'

Ten minutes after the evacuation the ship sank. It was in a place called the Black Deep, near Long Sand Bank. Only the top of its 160-foot transmission mast could be seen above the waves.

The same year as the sinking of the *Mi Amigo*, a film was made about Coxswain-Mechanic Charlie Bowry and the Sheerness lifeboat. It's called *The Making of a Crew*. You can watch it on YouTube. It was Charlie Bowry's son, the musician, who posted it there.

The film shows the coxswain training his crew in a number of exercises – man overboard, rescue by breeches buoy, first aid, emergency steering. By the time the film was made, the lifeboat station had only been in operation for ten years. Sheerness being a commercial port rather than a fishing village, there was, we're told, little sense of community round the lifeboat. In a fishing village, a lifeboat crew would most often be called out to save fellow-fishermen, but here nearly all the crew worked in the docks.

But for some there was a more personal stake. One of the volunteers, a carefully spoken man in his forties, with the manner more of a bank clerk than a seasoned mariner, tells us why he does it. 'I lost a brother in the water just up the river. I know – it doesn't matter if there'd been a hundred boats in the area – I know he'd never have been saved. But I've often felt that if anything of that nature happens again, perhaps it could be prevented.'

A younger volunteer says, 'If you sat down and thought about it you wouldn't do it at all. But it's one of those things. You do it, and then you stick to it.'

There are a number of scenes in the film showing the crew relaxing in a pub. It doesn't look like Old House at Home, and the coxswain is sitting supping his pint, not serving behind the bar.

They tell stories.

There was one night in really rough weather. It was 'snowing and sleeting and blowing' the coxswain says. One of the crew misjudged the wind and threw up over one of his shipmates. The coxswain didn't see this bit. 'The bit I see woz 'im,' he nods his head at one of his companions, 'throwin' buckets of water over 'im,' nodding at another companion, *at half past three in the mornin'.*'

The first man says, 'Tidied 'im up a little bit.'

'They gotta be in love,' says the coxswain, deadpan. 'Ain't they?'

There was another time, again at night, but this time in fog not storm, when the lifeboat went to investigate an apparently abandoned yacht. 'I thought,' says the coxswain, ''ello they've gone over the side 'ere. Cabin door's open. Moon's just coming through the mist now. Beginning to clear a bit, whole yacht's swaying about nice and calm. Come up alongside, lugged the big searchlight off the canopy, shone it down through the door. This fella and his girl sat bolt upright, naked as the day they woz born. Well she had these little red socks on, see. "Can you help me, I am lost," says this bloke. *He* says [gesturing to one of his companions], "You look like you're doing all right." Turned out they woz a Belgian couple. He says to me, "I am bumping everything in the Thames Estuary, and I am stopping here."' The lifeboat ended up towing the yacht through the night to safer waters, while the couple slept below. The Belgian man sent the crew a Christmas present every year after that.

Sometimes the mood is more thoughtful. 'The sea's a funny thing,' says the coxswain, puffing on his pipe. 'You never beat it. You can go out there and have a go. But all the time you know you're not the winner. And at any moment the sea decides, it'll just smack you back on your beam-ends, and you know who's boss.'

There are occasions when the lifeboat takes out a family so they can scatter the ashes of a loved one. After a short service, a wreath will be thrown onto the water. The lifeboat goes round in circles for a while to give the family time.

'I don't go to church,' says the coxswain. 'What would I want to go to church for? If I want to speak to God I can do it anywhere. I know he's there helping me when I'm out there. I truly believe that, otherwise I wouldn't do it, I wouldn't waste me breath. I might be having a quick few words in the wheelhouse when the lads are outside making things ready. Just to make sure he's about. I think it helps you through the nasty bits.'

However nasty the nasty bits, there were also many moments of heroism. The Sheerness lifeboat saved 200 lives in its first ten years. Charlie Bowry was awarded the RNLI silver medal for his role in the *Mi Amigo* rescue, having manoeuvred his boat along-side the sinking ship thirteen times in stormy waters. He'd already, in 1976, won the bronze medal after rescuing five crew from a yacht grounded on the West Barrow Bank in a southwest-erly gale.

At the end of the documentary, we see a fresh-faced, dark-haired boy kneeling on a stool, steering the lifeboat under the watchful eyes of the coxswain. The boy might be ten or eleven, perhaps even younger. 'Every man hopes his son is going to follow in his footsteps,' Charlie Bowry says. 'You can never be one hundred per cent certain of that until they leave school and they're standing on the boat there with you. Because they develop minds of their own, different ways they want to go.' He says his son was baptised on the boat, that the boat bonds the family together. One senses by 'family' he means not only his own flesh and blood, but also the crew. The boat is an extension of himself. 'You can *feel* the boat. A lot of people won't under-stand it. But you can *feel* the boat getting hurt. It's bonded us all together.'

That was all nearly forty years ago. The coxswain is now dead – otherwise his ghost wouldn't now be keeping an eye on the landlady in the small hours of the morning.

Is the young lad at the end of the film the grey-haired musician I met in the pub? He told me he was born in the 1960s, so it *could* just be him. I thought of my own father, of myself at that age.

The span of years appears impossible, an unbuilt bridge over an abyss as unfathomable as the Black Deep.

~

I finished my steak and kidney pudding and my beer, thanked the company, and turned to go.

'Just follow the path by the shore and you'll be in Bluetown in no time,' said the musician.

The afternoon was bright, blue-skied. The cold north wind carried the cries of gulls and the mournful call of a curlew, soon to leave its winter home on the coast for some upland moor. The way was clear, signposted as a public footpath. It followed an esplanade parallel to the shore. On the inland side was a concrete sea wall, curved above me like a breaking wave, dotted with patches of yellow lichen. On the other side the ebbing tide had uncovered mudflats. Beyond, yachts were moored in the mouth of the Swale. Further away, I could see across the Medway to the cranes in the docks on the Isle of Grain, great girder-legged things like the Martian killing machines in *The War of the Worlds*. Next to them were piles of pale gravel. Further on there were storage tanks, strings of pylons and a three-chimneyed industrial installation of some kind. On the Sheppey side, beyond the sea wall and a high barbed-wire fence, there was what looked like a chemical works. Ahead of me stood four wind turbines, and beyond them was the red funnel of a cargo ship in its berth at Sheerness docks. On the shore, the sharp posts of broken groynes, draped in black seaweed, jutted up into the sky like rotten teeth.

I was not allowed to follow the shore for long. A tall fence of metal spikes barred the way, and the path followed it inland, sandwiched between the fence and the sea wall. At regular intervals there were red-and-white signs declaring:

RESTRICTED AREA
PASS HOLDERS ONLY

The Isle of Sheppey

On the other side of the fence, between myself and the sea, was a vast car park, crammed with thousands upon thousands of imported vehicles, all still without plates. The asphalt, divided by white lines into innumerable parking bays, seemed to stretch for miles. There were no human beings visible anywhere. Instead there were rows and rows of Volkswagens and Vauxhalls, Audis and Citroëns, saloons, hatchbacks, estates, people carriers, builder's vans, plumber's vans, electrician's vans. Most were white, some black, a few silver. On the other, inland, side of the path, the sea wall was now topped with a barbed-wire fence, and bore such inscriptions as BEX ♥ TREV 07. Perhaps ten years before Bex had given Trev her heart here. Or something Trev might have been more eager to take. Wedged between the car park and the chemical works, the place wasn't conducive to romance.

The path itself – the public right of way – soon crossed some kind of channel or drainage ditch. The footbridge was completely enclosed in mesh – not only on each side, but along the top, forming a tunnel. It was the sort of place I imagined you might find in a slaughterhouse. For a moment I felt like a calf driven towards a bolt through the brain. 'Danger,' warned a sign. 'Protected by ultra-barb razor wire.'

Further on, on the other side of the public road that now ran parallel to my path, stood a massive industrial shed, walled in corrugated iron, painted in various shades of blue, smeared with rust. The lower parts of the walls had been stripped away. This was part of the Sheerness steel rolling mills, once run by

Thamesteel. Thamesteel had gone bust in 2012, resulting in the loss of 400 jobs. There are plans underway to reopen the plant, and to employ 100 workers, so what looked like demolition was in fact renovation. The site was once much more extensive, covering the acres now used to store imported cars.

~

Bluetown, the old part of Sheerness, forms a small triangle, separated from what was the naval dockyard by a blank brick wall, perhaps twenty feet high. It looks like the sort of wall once built round Victorian prisons. Today, Bluetown's a run-down area. The window frames of the terraced houses are flaking, the windows themselves opaque with greying net curtains. Most of the shop fronts are decayed and abandoned. There are still pubs – the Jolly Sailor, the Lord Nelson, the Royal Fountain Hotel, the Red Lion, the Albion Bar – but some are closed. There's now only a fraction of the thirty or so pubs that were open for business here in the nineteenth century. There is no outward evidence of the brothels that Baz had told me once thrived here, but there is a shop called Skorpio with painted-out windows. It offers 'adult fun', rated R18. As well as brothels, Bluetown once boasted a pleasure pier, but it closed more than sixty years ago. There was also a chapel, built by the dockyard workers in 1785 in their own time and at their own expense. John Wesley had already preached in Sheerness, mentioning the 'large and serious congregation'. There used to be a theatre, the Criterion, but it never recovered from its pounding by German Gotha bombers on 5 June 1917. On the gable end of one house there was an old ad, inscribed in plaster, for 'Budden and Biggs Body Building Beverages'. A more recent fly poster advertised a wrestling night, 'great fun for all the family'. There was also a sign for a boxing club, the Sheppey Academy of Excellence.

Although the Royal Navy dockyard closed more than half a century ago, the commercial docks are still busy. The camaraderie on Charlie Bowry's lifeboat, manned mostly by men who worked in the docks, seems to have been widespread. Another YouTube video, *Life in Sheerness Docks*, is a loving compilation of archive material from the 1960s, '70s and '80s. We see photograph after photograph of numbered gangs of dockers, standing in lines for the camera, to an elegiac soundtrack featuring such standards as Elgar's 'Nimrod', with added drumbeat and synth. As the music turns on an endless loop, each docker is named – for example, '27 Gang' comprises Brian, Colin, Grant, Noel, Terry and Gordon. Other men are just supplied with a surname and an initial. Some are given nicknames such as 'Nutty', 'Nobby' or 'Meatball'. They are all white men, nearly all with English names, mostly dressed in blue overalls. Some wear red rubber work gloves. The only women are those employed in the canteen. There are forklift trucks, mobile cranes, low-loaders, giant sheds. And ships. Some are passenger ships: from 1974 to 1994 the Olau Line ran a ferry service between Sheerness and Vlissingen in the Netherlands. But most are cargo carriers – *Baco-Liner 1*, *Maple Ace*, *Golf*. We see men unloading sacks of cocoa beans from Ghana and boxes of bananas from the West Indies. A ship called the *Kibishio Maru* is being loaded with explosives. Another ship, labelled 'Bomb Boat 1983', contains pallet after pallet of what look like artillery shells. No one seems to have stopped the dockers taking photographs of Britain's arms trade in action; and no one appears to have recorded where these exports were heading.

~

I started to walk down the road that led towards the commercial docks and the lifeboat station, but was brought up short by another red notice advising me I was entering a restricted area, and that my unauthorised entry would open me up to prosecution. I decided to walk eastward along the beach instead.

Looking back to the forbidden zone, I caught glimpses of giant gantries and three skeletal concrete forts, facing off threats from the sea. I was now on the north shore of Sheppey. Across the

Thames Estuary, beyond a distant Tilbury-bound container ship, stood the Essex shore and the high-rise blocks of Southend. The beach along this shore of Sheppey is shingle, with patches of sand, protected from longshore drift by groyne after groyne, their lines of posts marching down into the sea. Although sunny, there was still a chill breeze, and only a few people were on the esplanade. An Asian woman, in long green dress, trousers and headscarf, was taking a smartphone photograph of her three identically attired daughters. I followed the arrow on the esplanade pointing me towards

AWARD
BATHING
BEACH

No one was bathing in this weather, but a few mothers had pushed buggies onto the shingle. A sign warned of 'hidden groynes at high tide', while another spoke of 'hidden rocks & sharp shells', and suggested that I wear footwear on the beach and in the sea. A third sign proclaimed 'Excellent bathing water quality'. At Neptune Jetty, a stubby block of concrete jutting out into the sea, I was alerted to 'hazardous conditions', and advised against swimming, jumping or diving. A solitary, solidly built angler, dressed in black, his shaved head tanning in the sun, was taking up position at the end of the jetty. I asked him what he was after. Skate, he said. I asked whether it was good for skate round here. 'Red hot', he said. 'Up to ten pounds.' The 'skate' he (and Baz) referred to was not the common skate. The common skate, which can weigh in excess of 250 lb and can have a 'wingspan' up to nine feet, is critically endangered. The skate fished off Sheppey would more likely be the much smaller thornback ray, also known as roker. They get their name from the horns and spikes that cover their bodies.

I asked whether he'd had any luck. He said he'd only just set up. He was going to be there until ten in the evening. The thornback ray only feeds after dark. I said I'd never have the patience, but wished him good luck.

As I turned round and made my way to the station at Sheerness to catch the train back to London, a huge grey, red and white cargo ship slipped out of the docks. It was the *Bothniaborg*, a state-of-the-art roll-on/roll-off carrier, specially strengthened against the sea ice of a Baltic winter. It had just unloaded its weekly cargo of lorries carrying paper from Sweden, and now it was heading back into the Thames Estuary loaded with empty lorries before traversing the North Sea. It would then make its way through the Skagerrak and the Kattegat, sail under the bridge linking Copenhagen and Malmö before entering the Baltic proper. Passing by the islands of Bornholm, Öland and Åland, it would sail northward up the east coast of Sweden into the Gulf of Bothnia to pick up its next load of paper-carrying lorries from Haraholmen, a port nudged up against the Arctic Circle, before returning to England.

The *Bothniaborg* and I left Sheppey in different directions. I boarded my small train off the island to meet the main line at Sittingbourne, while the sleek freighter carved its way east to the Baltic. I thought of the skate the angler was after. As soon as the sun set they would be nosing the shingle beneath Neptune Jetty for shellfish and crabs. I thought of the spectral coxswain anxiously wandering his cellars in the depths of the night. I thought of the musician listening for the storm that would summon the ghost of his father back out to sea. Such a storm might prove powerful enough to crunch metal on crucial metal in the broken remains of the *Richard Montgomery*, setting off an unimaginable cataclysm.

But away from the time bomb ticking on the sunken sands offshore, away from the docks of Sheerness linking the island to the rest of the world, Sheppey is still the *Scepeig* of the Anglo-Saxons, 'the island where sheep are kept'. To this day, flocks still graze Sheppey's marshy pastures, oblivious of the threat lurking out at sea, oblivious even of the harriers quartering the ground, drifting over fields and reed beds, eyes ever alert, not for lambs, but for mice and frogs and fish. And all the time the tide, as it has always done, maintains its rhythm; and with the tide the island slowly pulses, growing and shrinking to the beat of the moon.

Waiting for the Ferryman

Wallasea Island

A sordid god: down from his hairy chin
A length of beard descends, uncombed, unclean;
His eyes, like hollow furnaces on fire;
A girdle, foul with grease, binds his obscene attire.

> – Virgil's description of Charon, the ferryman of the
> Underworld, in *The Aeneid*, translated by John Dryden

To get to an island you need a bridge, a low tide or a boat. There is a road bridge onto Wallasea Island, but it's a long way from anywhere. And the road's flooded for hours at a time when there's a high spring tide. It's easier to travel by train to Burnham and board the small passenger ferry that cuts across the estuary of the River Crouch. The ferry runs on demand. You just phone Jon the ferryman and he'll meet you at the quay.

A ferry crossing involves a transition. It might be a shift from the certainty and solidity of the mainland to the provisionality of an island, where land may at any moment become water, and vice versa. Or the transition may be, in certain mythologies, a journey between the realm of the living and the realm of the dead. The Greeks believed the souls of the departed were ferried across the River Styx to Hades by the gloomy figure of Charon. In recounting Aeneas's descent to the Underworld, Virgil gives a chilling description of the ferryman who controls this dreary crossing.

He's dressed in greasy rags, Virgil tells us, and from his chin dangles a squalid straggle of matted beard. His eyes burn like furnaces in his head.

Charon only allows those who have been buried or burnt on a funeral pyre to board his craft. Even those who have been given the proper rites have to pay Charon his fare, and it was thus the custom to place a coin under the tongue of the recently deceased. Those shades who cannot pay, or those whose corpses have been abandoned on the field of battle, are doomed to haunt the near shore for a hundred years, yearning for their destination, hoping for Elysium.

On the train from Liverpool Street, rattling eastward into the grey, empty skies of Essex, I found myself in the company of the dead, or at least of the despairing, which is a kind of dying. On the seat in front of me, unseen behind its high back but clearly heard, a Glaswegian builder was on his mobile. There was a crisis on one of his sites, somewhere near Trafalgar Square. Workmen had punctured an old lead pipe and water was spurting everywhere. 'Fucking all over the place,' he said, and groaned. 'Send me a picture,' he told the man on the other end of the line. 'I'll get on to it. Fucking hell.' More phone calls, more swearing, as he desperately tried to fix things from a distance. Then he phoned his mother. 'Mum, I'm fucked. Don't know what's wrong with me. Drinking every night, never used to, fucking not right, can't sleep. I'm fucked, completely fucked. Something's got to change. I've got to get out.' He went on like this for some time, baring his soul for all to hear, although the carriage was nearly empty. Eventually he drew the conversation to an end. 'Mum, just phone me when you've some good news. I can't take any more. Fucking had it.' He was a long way from home. His life was unravelling. I wondered if, in his disturbed, angry mind, he was looking into the burning eyes of the ferryman, begging for a ride across the dark water.

Jon, the flesh-and-blood boatman of Burnham-on-Crouch, was a far cry from Charon, the 'foul and terrible' ferryman sculling his rusty hull across the Styx. Jon was neatly dressed in red jacket and shades, clean-shaven, weather-beaten, friendly. He lent me his hand to help me aboard his RIB, a smart blue SeaRover 18. I was the only passenger. I handed him four pound coins for my fare. If I'd been a babe in arms I would have travelled for free, if a dog I'd have gone for half price. The big outboard roared into life, and Jon steered the RIB in a leisurely, swooping curve back up the river towards the island. The crossing only takes ten minutes. By car it's an hour's drive from Burnham to Wallasea.

I asked Jon whether he was back and forth on the river all day. He said only when someone phoned him, he couldn't afford to keep running without passengers. He owned the boat jointly with his brother, and they worked alternate weeks, from Easter until the end of September, Wednesdays excluded. They did manage to make a small profit, he said. Not a lot, but it was handy pocket money to top up their pensions.

As we approached the island I saw through the gloom that the water off the north shore was crammed with wreckage, a tangle of rafts of wood and rusted iron, grass growing from the cracks. Some of the sections were tilted up at crazy angles, some were linked to others by chains. Jon explained that during the war they'd formed an anti-submarine barrier in the Thames. After the war they'd been towed round to the Crouch and used as pontoons for yacht moorings. 'Then in that big storm – 1978 was it? – they all broke away and ended up on the Burnham shore with all the boats attached. That was a very expensive junkyard.' Now the wreckage had been dumped on the mudflats of Wallasea. The Wallasea marina had been rebuilt, and now rows of gin palaces with names like *Bellarina* and *Don Papa* were moored along the replacement pontoons.

I arranged to phone Jon before 4.30 p.m. for the return trip. That would be his last sailing back to Burnham. I walked up the floating jetty and set off eastward along the dyke by the shore. On top of this dyke runs the only footpath on Wallasea. It was high tide, and the land on my right was lower than the sea on my left.

The Ordnance Survey marks not a single contour on the entire island. The highest spot height is five metres, and even that feels like an exaggeration.

In the past, the low-lying marshy islands of eastern Essex were not only subject to regular flooding, but also notoriously malarial. Locals built up some kind of immunity to the 'marsh ague', but if island farmers took wives from the mainland, the women were often dead within the year. Three centuries ago, Daniel Defoe observed that hereabouts 'it was very frequent to meet with men that had had from five or six, to fourteen or fifteen wives'.

Among the native women of Wallasea, legend has it, was a witch called Mother Redcap, who lived in a decrepit farmhouse called Tyle Barn, also known as the Devil's House. Tyle Barn was located on the south side of the island, overlooking the stretch of the River Roach still known as Devil's Reach. Tyle Barn was said to have been built on the instructions of the Devil himself, who threw a beam in the air and told some labourers to build a house wherever the beam landed. A more likely explanation is that the house was once lived in by a man called Davill, although the legend insists that it was haunted by Mother Redcap's familiar, a creature of uncertain shape whose heavy wingbeats could be heard up under the roof. Mother Redcap was also said to live on Foulness Island, on the other side of Devil's Reach. Tyle Barn itself was bombed in the Second World War by the Germans, who presumably mistook it for one of the many military installations on Foulness, which has been used as a weapons-testing area by the armed forces since 1848. Public access is severely limited. The ruins of Tyle Barn were swept away in the great floods of 1953.

Centuries before this, much of the island's marshland had been drained for farming, but now most of Wallasea is owned by the RSPB, who are in the process of restoring large areas of wetland – marsh, mudflats, lagoons, artificial islands. These provide habitats for birds such as avocets, brent geese, little egrets and black-tailed godwits, and breeding grounds for fish and even dolphins. When completed, Wallasea will be the largest man-made marine wetland area in the UK, intended to replace the extensive marshlands drained to make way for the expansion of the ports at Felixstowe and Sheerness.

In the distance, as I walked on the dyke along the northern shore, I could make out big industrial conveyor belts. These had been used for bringing ashore ship-borne spoil for the wetland restoration. Huge amounts of spoil had been acquired from the excavations for London's Crossrail scheme – the same excavations that turned up the skeletons of 3,300 plague victims, now reburied on Canvey Island, just a few miles away to the southwest. Jon told me that work had come to a halt on Wallasea as all the spoil from the Crossrail excavations had been used up. He said they'd tried to get some from London Gateway, just west of Canvey, where they've dredged a new container port, but that fell through. Now three-fifths of the wetland restoration work has been completed, but the rest is on hold, Jon thought. So the western part of the island is still covered with flat farmland. The remaining fields, more like an endless, unfenced prairie, are edged by wide, reed-filled ditches.

On a wooden footbridge across one of these ditches perched two herons, each on its own post a few feet apart. In the reeds below them I could make out the white shadow of an egret. As I approached, the herons and the egret slowly flapped into the air. Perhaps it had been one of these great birds that had haunted Tyle Barn, its wingbeats breaking the silence, filling the air with an uncanny chill.

Further on, I came to the new lagoons of the RSPB reserve. Gulls in great numbers were nesting on the islands, making their usual din. Then I heard a persistent peep peep. I thought at first it must be an oystercatcher, but the pitch was too high. Then I saw it: the white of an avocet, its wings barred with black. It was soon joined by another. They wheeled about for a while in fast, nervous circles, clearly alarmed, but eventually settled. I thought of the souls of the unburied dead, destined to haunt the wrong bank of the Styx for a hundred years.

On the east shore of Wallasea, overlooking the stretch of the River Roach called Brankfleet, the path came to an end. The dyke stopped, and a creek barred the way. There was a bird hide, where I sheltered from the east wind to eat my sandwiches. I looked across the wide, sluggish river to Foulness. There was no bridge, no sandbar, no ferryman to take me across. My journey had come to an end.

Then the peace I shared with the breeze and the songs of larks was broken by a loud bang. Over on Foulness, the MoD's forbidden island, beyond watchtowers and radio masts, a plume of black smoke rose into the sky. Britain's weapons were being tested. As I looked across the water, I wondered whether someone on the other side was watching me.

~

It was a long walk back. I'd decided to make a sort of circle, so left the path on the dyke along the north shore to follow a single-track concrete road through the middle of the island, surrounded by a green sea of flat wheat fields. The giant barn of Grapnells Farm, visible from all over the island, grew no closer. I had to keep up a good pace if I was to make the last ferry. My feet ached on the hard concrete, and pains began to shoot up my shins. A hot, pale sun broke through the dark grey cloud and beat down on my forehead. I looked for a shortcut back to the north shore through the Essex Yacht Marina, but there was no public access. The entrance to the timber yard next door, signposted 'Baltic Distribution', was similarly discouraging. I ended up crossing the road onto the mainland at the far western end of the island. On one side, Paglesham Creek was more or less dry, on the other side the tide had ebbed out of the salt marshes of Lion Creek. I walked along the top of a grassy dyke, parallel to Riverside Village Holiday Park, with its rows of fixed-pitch caravans nestled among the willows. Here and there a Union Jack or a Cross of St George fluttered in the breeze.

Beyond the caravan park, there's a 1960s Tudorbethan pub called the Creeksea Ferry Inn. It was closed. In fact, it was derelict, boarded up. This must have been the successor to the old Creaksea Ferry Inn whose landlady had survived the night of the great North Sea flood of 1953 by hanging onto the top of a door for nearly seven hours, while the water washed around her neck. Two of her customers also survived, perched on the roof through the bitter winter night. A third had determined to swim off in the dark for help, a torch in his hand. He was later found drowned, still clutching the torch. Several weeks later, another body was found, that of the island's auxiliary postman. His corpse was discovered in a ditch behind the sea wall.

On a slipway back in the here and now, three anglers were preparing their rods, accompanied by a very large Staffordshire bull terrier. At the small dock called Baltic Wharf, timber was being unloaded from a cargo vessel, the *Mosvik*, flying the Red Ensign but registered in Antigua Barbuda. I tracked it online. A couple of days later it was traversing the Kiel Canal back into the Baltic to fetch more timber. Apparently a lot of Russian ships also offload at Wallasea, sometimes landing steel.

Jon the ferryman was waiting for me at the floating jetty, and helped me aboard. Again, I was his only passenger. I sat down near the bow. 'Better sit back some more,' Jon said. 'We've got wind against tide. Could be choppy. We might get a bit of a slapping.' We did. As we pulled out into the middle of the channel a small wave hit the bow and splashed over me. 'Sorry about that,' said Jon. I reassured him, as I tasted salt on my lips, that it added to the island experience. I was, above anything else, glad to be returning to *terra firma* from the empty wet flatlands of Wallasea, a place more suited to fish and fowl than people.

And so I made my transition back to the mainland, a place where the living can get on with their humdrum lives. But as we crossed back over the estuary, I remembered the slow heavy wing-beats of the herons and the egret; the anxious, quick flight of the avocets. For a moment the shades of the stranded dead flitted across my mind. But in another moment they had settled back into the obscure dark.

Into the Heart of Darkness

The Isle of Dogs

'And this also,' said Marlow suddenly, 'has been one of the dark places of the earth.'

– Joseph Conrad, *Heart of Darkness* (1899)

The Isle of Dogs: the very name evokes a brutish kind of place, a dead end abandoned by humans, drained of any last drop of pity. It brings to mind a dystopia of deserted, ruined streets, a nightscape where feral packs snarl and sniff for bones around the rims of oil-slick puddles.

The Isle of Dogs is not in fact an island. It's a peninsula – a word and concept formed from the Latin *paene*, 'almost', and *insula*, 'island'. This quasi-island juts into the Thames opposite Greenwich, forming the biggest loop in that winding river. It is isolated on three sides by Limehouse Reach, Greenwich Reach and Blackwall Reach; the Thames from here to the sea has many names. On the fourth (northern) side, the Isle of Dogs was in the past cut off at the neck from the mainland by marshes. For many centuries, the whole area was, in fact, marshland. Locals have always referred to it as 'the Island'. You can reach it by road, or Tube, or via the Docklands Light Railway. Or you can walk to it under the Thames through the Greenwich Foot Tunnel, emerging at Island Gardens, or by a short ferry ride across the river from Rotherhithe.

Geologically, the peninsula consists of alluvial silt, deposited by the Thames over the millennia on layers of mud and clay. The marshes were drained in the thirteenth century, but inundated by a great flood in 1488, reverting to marshland until the seventeenth century when Dutch engineers again drained the area. Prior to this, the peninsula was a notoriously unhealthy swamp, the repository of much of the sewage flowing downriver from Westminster and the City. As such, in the satirical 1605 comedy *Eastward Hoe* (a collaboration between Ben Jonson, George Chapman and John Marston), the place provides an appropriate landfall for Quicksilver and Sir Petronel Flash, a pair of conmen who are hoping to escape to Virginia with their ill-gotten gains. Instead, they end up shipwrecked on the Isle of Dogs, at a place called Cuckold's Haven. The play so displeased King James I with its anti-Scottish references that Jonson and Chapman spent a month or two in prison, and only just avoided having their ears and noses slit. Earlier, in 1597, Jonson had got into trouble over a satirical comedy called *The Isle of Dogs*, co-written with Thomas Nashe. The authorities judged the play so 'lewd' and full of 'slanderous matter' that it was immediately suppressed. Arrest warrants were issued for the authors. Jonson spent some weeks in prison, but Nashe was out of London at the time and escaped punishment. No copy of the play survives.

By the middle of the Victorian era, the Isle of Dogs was living up to the connotations of its name. Thomas Wright, in *Some Habits and Customs of the Working Classes* (1867), describes the 'tumble-down buildings, stagnant ditches and tracts of marshy, rubbish-filled waste ground', while in 1911 George R. Sims calls it 'Desolation-Land', and writes of 'a black fringe of grim wharves and towering chimneys'.* A decade later, in T. S. Eliot's *The Waste Land*, the Thames 'sweats / Oil and tar' while logs drift 'Down Greenwich Reach / Past the Isle of Dogs'. When the poem was published in 1922, the Isle of Dogs must have seemed

---

* George R. Sims, *In Limehouse and the Isle of Dogs* (1911)

to a cultural pessimist like Eliot to embody the worst of the modern world. It is unlikely he ever visited the place.

The name itself has long been a puzzle. In the Middle Ages the Isle of Dogs wasn't the Isle of Dogs at all, but was known as Stebunhethe Marsh. Stebunhethe (first recorded *c.*1000 as Stybbanhythe) is an old form of the name Stepney, which in medieval times was a large parish extending from the City of London in the west to the River Lea in the east. The present name was first recorded in 1520. But why *Dogs*? It has been suggested that it was an irreverent imitation of the Isle of *Man*, perhaps in reference to the royal hunting dogs that may or may not have been kennelled here, across the river from the royal palace (or hunting lodge) at Greenwich. Some say the relevant king was Edward III, others mention Henry VIII, but there is no firm evidence for any of this. Shakier still is the suggestion that the dogs concerned were dead ones washed up on the western shore of the peninsula. On even less firm ground teeter the rival theories that the name was originally 'Isle of Docks' or 'Isle of Ducks'. The former is implausible: although goods had been landed on the peninsula for centuries, there were no properly constructed docks here until the West India Docks were opened in 1802. The latter theory, that the name derives from 'Isle of Ducks', would allude to the wealth of wildfowl once found here. However, linguistically as well as zoologically, ducks rarely evolve into dogs. A further suggestion is that the Isle of Dogs alludes to Gran Canaria ('great [island] of dogs') in the Canary Islands, whose famous wines might well have been landed here. Somewhat confusingly, the huge modern financial centre known as Canary Wharf is named after the quay and warehouse, originally No. 32 Berth on the North Dock in the West India Docks, where fruit from the Mediterranean and the Canary Islands was landed. But No. 32 Berth was not built until 1936, and was named in allusion both to the Canaries and to the Isle of Dogs.

A map of 1747 shows the Isle of Dogs as a pattern of fields and lanes, almost severed from the mainland by a slit of water called

Poplar Cut, roughly where South Dock is today. (The district of Poplar, then a hamlet, occupies the northern end of the Isle of Dogs.) These fields were used to fatten livestock for the markets of London, and also supplied meat to the Royal Navy. Along the western shore, dotted with windmills, the 1747 map marks 'Marsh Wall' – hence the later name 'Millwall' for the settlement that grew up here in the nineteenth century to house new populations of shipbuilders and dockers. By the end of the eighteenth century, the rapid expansion of Britain's overseas trade meant that the docks further up the Thames, closer to the City, had reached capacity. The West India Docks, at the neck of the peninsula, were opened in 1802, followed by the East India Docks in Blackwall in 1806 and Millwall Dock in 1868 (the spoil from the excavation of this last dock was dumped at a place still known as Mudchute; locals long complained of the smell). The West India Docks were connected by locks to the Thames at both western and eastern ends, and so, for a while at least, the Isle of Dogs could have been described as a genuine island. At the same time, it became the heart of a vast commercial empire, its veins and arteries sprawling across the oceans of the world.

At the northwest end of the West India Docks there is a large plaque made of slabs of Portland stone set into a high brick wall. The wall was built to protect the docks from thieving. On this plaque is inscribed a self-congratulatory celebration of the new undertaking:

Of this Range of BUILDINGS
Constructed together with the Adjacent DOCKS, At the
Expence of public spirited Individuals
Under the Sanction of a provident Legislature,
And with the liberal Co-operation of the Corporate Body of
the CITY of LONDON,
For the distinct Purpose
Of complete SECURITY and ample ACCOMMODATION
(hitherto not afforded)

To the SHIPPING and PRODUCE of the WEST INDIES at this
wealthy PORT.

THE FIRST STONE WAS LAID

On Saturday the Twelfth Day of July, A.D. 1800,

BY THE CONCURRING HANDS OF

THE RIGHT HONOURABLE LORD LOUGHBOROUGH,

LORD HIGH CHANCELLOR OF GREAT BRITAIN,

THE RIGHT HONOURABLE WILLIAM PITT

FIRST LORD COMMISSIONER OF HIS MAJESTY'S TREASURY;

AND CHANCELLOR OF HIS MAJESTY'S EXCHEQUER,

GEORGE HIBBERT, ESQ, THE CHAIRMAN, AND ROBERT

MILLIGAN, ESQ, THE DEPUTY CHAIRMAN

OF THE WEST INDIA DOCK COMPANY:

The two former conspicuous in the Band Of those illustrious
Statesmen

Who in either House of Parliament have been zealous to promote

The two latter distinguished among those chosen to direct

AN UNDERTAKING

Which, under the favour of GOD, shall contribute

STABILITY, INCREASE, and ORNAMENT

TO

BRITISH COMMERCE.

Commerce was now not only Britain's top priority but its
transcendent glory, sanctioned by a benign Parliament, encouraged
by the nabobs of the City, blessed by His Majesty's Government
and looked on favourably by the Almighty Himself. What these
old stones fail to mention was that the principal cargo landed at
the West India Docks was to be sugar – then still grown, harvested,
processed and sold on the back of slave labour.

Although Britain was to outlaw the slave trade five years after
the West India Docks opened, slavery itself continued to be

permitted in the British Empire until 1833. Robert Milligan, the deputy chairman of the West India Dock Company eulogised in the memorial, and the moving power behind the development, had grown up on his family's Jamaica sugar plantations. At his death in 1809 he was the owner of 526 slaves. On the north side of the West India Docks, outside No. 1 Warehouse, a building that now houses the Museum of London Docklands, stands a bronze statue of Milligan, erected after his death. Bewigged, breeched and comfortably paunched, Milligan lords it on his plinth, on the flank of which is fixed a bas-relief depicting a woman and her children laying the fruits of empire at the feet of a haughty, androgynous Britannia. She sits on the back of a cross-looking lion, holding it in check with a hand on its mane.

Neither the plaque celebrating the foundation of the docks nor Milligan's memorial give any inkling of the cruelty woven into the wealth that sugar brought. For that, one has to turn to the scarce accounts left by the slaves themselves, such as *The History of Mary Prince*, published in 1831, while slavery was still legal in the Empire. Born into bondage in Bermuda, Mary Prince was sold several times. Arriving at the house of a new master, she recalled that 'The stones and the timber were the best things in it; they were not so hard as the hearts of the owners.' Of one mistress, Prince wrote, 'She caused me to know the exact difference between the smart of the rope, the cart-whip, and the cow-skin, when applied to my body by her own cruel hand.' Although Milligan's statue was erected in 1813, it was not until 2007 that a plaque to Mary Prince was unveiled in Bloomsbury, London, where she had lived towards the end of her life. (In June 2020, three years after I first encountered Milligan on his plinth, and following widespread protests by the Black Lives Matter movement, his statue was removed.)

The self-deception of those who profited from slavery is starkly illuminated by the following passage from a letter written in the 1770s by John Pinner, a young man from Bristol who owned sugar plantations on the island of Nevis: 'I can assure you I was shocked at the first appearance of human flesh exposed for sale. But surely God ordained 'em for the use and benefit of us. Otherwise, his divine Will would have made itself manifest by some particular or token.'

In contrast to this casuistry, others found their consciences more consistently pricked. In the previous century, Thomas Tryon, an English merchant, hat-maker and vegetarian, had travelled to Barbados and had been horrified by the treatment of the slaves. Here he describes working conditions in a sugar mill:

> . . . the servants [i.e. slaves] night and day stand in great boiling houses, where there are six or seven large coppers or furnaces kept perpetually boiling; and from which with heavy ladles and scummers they skim off the excrementious parts of the canes; till it comes to its perfection and cleanness, while others as stokers broil as it were alive, in managing the fires . . .

Sugar was indeed a cruel trade. In the West India Docks, the area between the transit sheds on the quayside where the sugar was unloaded and the warehouses where it was stored became known as 'Blood Alley' or the 'Blood Hole'. The dockers had to carry bags weighing a crippling two or three hundredweight (100–150 kg) on their backs, and the damp, gritty sugar left their skins cracked and bleeding. Dave Matthews, one of my guides at the Museum of Docklands the day I visited, told me his father had been a docker here. He remembers that when his dad came home from work his mother had to pour water onto his shirt, then carefully peel it off, revealing the sharp crystals of sugar embedded in the skin of his back. The next week he might be carrying bags of cement, and the lime would burn into the cuts left by the sugar. When the cargo was flour, the men would drink so much water to quench their thirst that the flour in their throats would turn to paste and they would choke. Dave's dad died aged sixty-one.

For some dock workers there were compensations. The coopers who maintained and repaired barrels of wines and spirits would sometimes draw off illicit samples, a procedure known as 'sucking the monkey'. Some dockers profited from the imports of expensive women's clothing, donning layers of silk under their working clothes to smuggle out their booty. Dave told me a story about a docker who would leave the docks every day with an empty wheelbarrow. The policeman who monitored the gate was suspicious. The docker was often searched, but nothing was ever found. After many years of this, the policeman retired. Meeting his suspect in the pub, he bought him a drink. The ex-constable told the docker he knew he was thieving, but always wondered how he did it. 'You can tell me now. I'm retired, you won't get nicked. So, what were you thieving all those years?'

'Wheelbarrows,' said the man.

Sugar wasn't the only cargo to be landed on the Isle of Dogs. At Millwall Dock they unloaded tomatoes and potatoes from the Netherlands, oranges and lemons from Spain, barrels of port wine from Portugal, dried fruit from the Mediterranean and Australia. From East Africa there was sisal, and from southern Asia thousands of bags of shellac, bales of jute and hemp, crates of peppers, sacks of spices. I gleaned all this from listening to one of the tapes stored in the audio archives of the Museum of London (parent of the Museum of Docklands). They hold a number of recordings of people who had lived and worked on the Isle of Dogs. The interviews were mostly conducted in the mid-1980s, after the docks had closed, and before the construction of the Canary Wharf development had begun. The interviewees are all long dead.

One of the voices is that of Doug Mason. After being demobbed in 1919 as a twenty-year-old, following four years' service in the army, Doug had worked in a clerical capacity at Millwall Dock until he retired in 1966. He called his line of work 'wharfinging'.

For manual work the docks relied on casual labour, up to 200 men a day per ship when a fruit boat came in. A bell would go at a quarter to eight at Millwall Gate, and the men would jostle and fight for position, hoping they'd get picked. The labour master would call out names from a list, then the chosen ones would go through into the yard and be allocated to different gangs. 'Those men were on their feet all day long, for sixteen bob [80p] a day. And they really worked. If they didn't work, they wouldn't be taken on the next morning.' Doug had no illusions about the good old days.

Still, he had immense pride in the work that he'd done. 'There's a lot in it. A *vast* amount of work in wharfinging. People don't realise, the different things you have to do, and things you shouldn't do.' The interviewer asks him to be more specific. 'Ah no,' he chuckles warningly. 'Tricks of the trade.' One frequent problem was maggots in the dried fruit. The boat would be sealed up and gas pumped in. The seals were opened after twenty-four hours to let the gas out, and after another twelve hours it would be safe for the men to go aboard to begin unloading.

For his first three years working at Millwall Dock, Doug was in local lodgings. Social life consisted of swimming, church, dances and night school, all on the Island. The locals – 'lovely people' – didn't 'go off' the Island. By 1926, after three years courting Bessie, Doug had saved up enough money to get married. By then he was on £4 5s (£4.25) a week, enough to set up a home in Bermondsey, a little way upriver on the other side of the Thames, close enough for Doug to cycle to work on the Island via the Greenwich Foot Tunnel. During the Second World War he and Bessie got bombed out three times. The last time was in 1944, 'the first morning of the rockets' (the V-1 flying bombs, also known as buzz bombs or doodlebugs). It was half past seven in the morning, and his wife was in bed, sick with cancer: 'I was in Greenwich Park when it dropped. When I got to the wharf, the manager said, "You'd better go home again, you've got trouble." It had blasted our roof right off.' He gives a hollow chuckle. 'Well, it's one of those things.' Two weeks later his wife was dead. Of the shock, he said.

London's docks, unloading vital foodstuffs from the convoys, were under tremendous pressure of work throughout the war, handling at any one time up to a thousand ships. Doug particularly remembers the fruit from Australia: 'Beautiful fruit. Sultanas, currants, raisins, dried apricots, peaches, nectarines . . .' Despite the temptations of the money to be made on the black market, Doug says there was very little pilfering. Everybody knew the police came down heavily on such activities. 'Very 'ot they were.'

Doug says Millwall Dock largely escaped bomb damage during the Blitz, although bombs fell very close by. The men often ignored the air-raid sirens so they could get on with vital work. Doug and some of the other clerks spent many of their nights on the Island as firewatchers.

As a whole, London's Docklands were badly hit. Another voice from the audio archive, Annie Pope – who was born in Cubitt Town in the southeast of the Island in 1909 and who lived there all her life – also has vivid memories of the Blitz. 'I never thought the Island would survive,' she says, 'because on that first Saturday when the planes come over, it was terrible.'

From the air, the Isle of Dogs – more specifically the loop of the Thames that winds round it, reflecting any light there is in the sky – provides a giant signpost pointing at itself. The Luftwaffe called this loop the *Zielraum* – the target area. That was where the docks were concentrated.

When on the afternoon of Saturday 7 September 1940 fleets of hundreds of bombers and fighters were seen flying up the Thames towards London, some casual watchers on the ground assumed they were RAF planes on exercise. Then the Ford factory at Dagenham was hit, and the huge gas works at Beckton, then the West India Docks. Streets of neighbouring houses were also hit. The bombers returned again after nightfall, their path marked by flames. They were to return again and again, for many more nights. Round the West India Docks people remembered the smells of burning spices, burning dust, burning tea, the night sky as bright as day. The streets round about ran with molten sugar. When it cooled to caramel, the sweet-starved local kids chipped away and gorged themselves.

'Everywhere, right the way round the Island, was in flames,' Annie Pope remembers. 'One mass of flames all the way round.' She pauses. 'Terrible when the war was on.' She and her family felt safe in their shelter when they heard the anti-aircraft battery on the nearby Mudchute opening up. But the guns failed to save the forty people killed and sixty injured on the evening of 19 March 1941, when the public shelter at Bullivant's Wharf received a direct hit. Annie quietly alludes to this tragedy, the biggest wartime disaster on the Isle of Dogs. Then she sighs. It is a sigh of resignation.

By 1944, with the advent of the V-1s, people were getting a little more blasé. Annie says that when the buzz bombs came over, that was always the time her daughter chose to have a bath. Her daughter would often have to run to the air-raid shelter wrapped in a towel.

Annie had lost out on her schooling during the First World War. They were so short of teachers that the girls were put to knitting socks for the troops, and learnt little. She went on to work at Burrell's colour factory on the Island, for 9s (45p) a week.

Once she was married, Annie was obliged to give up work. But with the onset of the Second World War she went back into employment, at Maconochies, making rations for the soldiers: 'Pickles, sauces, meat puddings, everything you could think of in the eating line. Sweets, marshmallows, everything you could think of in the chocolate line. You'd get tired of eating them. You could eat as many as you liked while you were in there, but you couldn't take any home.'

She was searched on many occasions. But she says she never tried to take anything out.

Annie's husband was a lighterman, a lighter being a type of flat-bottomed barge used to ferry goods between ships and riverside wharves. He had to work with the tide. Sometimes he didn't get home until ten o'clock in the evening, sometimes he had to work all night. Before the unions fought the issue, he'd be expected to work all the next day as well.

Eventually he became an under-foreman at Millwall Dock. She was never allowed to meet him there: 'They used to have police on the gates. Those police knew every day of their working lives who was going in and where they were going.'

Both dock labouring and lighterage were insecure. Annie remembers that when her husband went out in the morning, neither of them knew whether he was going to work that day. If there was no work he had to sign on for three days before he received any unemployment benefit. And that didn't go far. They had to pay for the doctor; and to have each of her children delivered she had to give the midwife 10s. If she went to the dentist to have a tooth extracted, she had to drop a shilling in the box. Her husband never made it to pension age. He died at the age of fifty-seven.

But what Annie remembers above all is the communal spirit on the Isle of Dogs. Although many came onto the Island each day to work, those who lived there had a strong, perhaps closed, sense of community. She remembers how during the war people looked out for each other. After every air raid, neighbours would call out to make sure everybody was ok. 'Very, very good neighbours, all helped one another.'

Despite this community spirit, two-thirds of the population left the Island during the war. Many didn't return.

~

In the end it wasn't the Luftwaffe that did for the London docks. It was containerisation. The old docks were simply not big enough. Nor was the Thames this far upriver deep or wide enough to deal with the huge new container ships. Instead, a specialised container port was built downriver, at Tilbury. The work of the traditional London docker was over. One operator of one vast container crane could load or unload a ship virtually single-handed, where before it would have needed scores of men. On the Isle of Dogs, the last of the docks closed in 1980.

Decline had set in long before. In protest at the perceived neglect of the Island by both local and central government, on 1 March 1970 a thirty-seven-year-old office worker and Labour councillor called Ted Johns issued a unilateral declaration of independence. For a couple of hours he and his supporters blockaded the two swing bridges that then provided the only road links to the Isle of Dogs. 'We can govern ourselves much better than they seem to be doing,' Johns declared. 'They have let the Island go to the dogs.' A thirty-member Island Council was established, with Johns as president of the new republic. However, within ten days, amidst mutual recriminations, the status quo had been restored.

The establishment in 1981 of the London Docklands Development Corporation by Michael Heseltine, Margaret Thatcher's environment secretary, turned out to have more concrete consequences. A new enterprise zone, free from local-authority planning control, was established in an area that encompassed the West India, East India and Millwall Docks. Beginning in 1987, part of this became the site of the Canary Wharf development, a project to create a vast new financial district to compete with the City of London, and a skyline to rival that of Manhattan. Symptomatic of the relaxation of planning conditions, firms were allowed to display their logos and names in large letters on the exteriors of their offices – something that was never tolerated in the City of London. So now the buildings are branded: KPMG, Citi, HSBC, Barclays, Credit Suisse, Bank of America, J. P. Morgan. Many of the buildings were designed by American architects. Glass, steel and concrete dominates, but there is also plenty of pastiche to appeal to faux-traditionalists, such as the pillared entrance and vaulted arcades of 10 Cabot Square. Dominating it all rises the Canary Wharf Tower, properly One Canada Square, a monolithic fifty-storey office block topped by a pyramid, completed in 1991. At 770 feet tall it was, until the completion of the Shard in 2012, the UK's tallest building. On its opening, its architect, the Argentinian-American César Pelli, described his creation, with an architect's typical modesty, as 'a portal to the sky', and 'a door to the infinite'.

Quarter of a century ago when I wore a suit and worked for a rather grand publisher in a posh part of town I boarded the Docklands Light Railway at Bank. It was to be my first visit to the Isle of Dogs. The DLR had been opened in 1987, a driverless train linking the City to a Dockland without docks. Alighting at Canary Wharf I found myself at the foot of One Canada Square. I'd seen the tower in the distance from the motorways of Kent and been hit by a laser beam projected from the pyramid down a street in Hackney near to where I lived at the time. Clutching my briefcase I entered the atrium at the foot of the tower. All about me were striped shirts and red braces, rosy plump or thin grey yuppie faces as purposeful and soulless as ants. You probably couldn't have distinguished me from the rest.

I joined the throng, found the general reception, stated my purpose, signed in, was issued with a pass and pointed to the lift. I'd arranged a meeting with the publications manager of the *Telegraph*, in the hope that some of the paper's journalists might have a few proposals for books I could commission. I'd signed a contract with one devil to get my job. Now I was to dine with another.

It wasn't quite like that. The publications manager was a kind woman, older and wiser than I was. We ate lunch alone in a private dining room on the eleventh floor. The room was on a corner of the tower and had floor-to-ceiling glass. We looked down the Thames out towards the Estuary as we talked over various ideas. It was a stunning view, with expensive new developments growing up all over Docklands. The river, once shunned by the metropolitan rich as a filthy sewer, a place of work not pleasure, was on its way to becoming their playground. Moneyed London had stopped turning its back on the Thames. Our meal and conversation concluded, I bought a rather charming book about the idiosyncrasies of the English weather.

It was not until 2017 that I returned to Canary Wharf. This time I was to go on a guided architectural tour led by the urban historian and architectural researcher, Mike Althorpe, who calls himself the London Ambler. The first thing Mike told us was that the tour could be stopped at any time. There were no rights of way in Canary Wharf, he said. Even the roads were private. We'd have to hope for the best. And we were to be careful about taking photographs. Security tended to be very touchy, particularly around entrances. The Canary Wharf Group owns over 20,000 square yards (1.81 hectares, to be precise) of apparently public space on the Isle of Dogs, including Cabot Square, Westferry Circus, West India Quays and Canada Square Park. It is a pattern that is increasingly seen across London, as cash-strapped local councils, unable to fund public spaces, make agreements with private owners who are allowed to draw up their own rules for these 'pseudo-public' spaces. The owners do not have to publish the rules, and hire private security firms to enforce them, whatever they might be. One thousand years ago, Norman barons did something similar when they turned the forests of England into their own private hunting grounds and hanged any peasant found in them, on suspicion of poaching. Vast areas of common land suffered a like fate during the enclosures that started in the Tudor period.

We began our tour on the shore of the Thames in the angle between Limehouse Basin and the Isle of Dogs, close to the places marked on the 1747 map as Limekiln Holes and Cuckold's Point. It was a hot lunchtime in June, and at the top of the steps leading down to the river a young woman in a bikini lay sunbathing, her office clothes abandoned at her side. We made our way round the shiny new financial district, walking in amongst imposing bank headquarters, frothing fountains, chunky pieces of newly commissioned sculpture. We trod the old quays, now home to a handful of billionaire's gin palaces with signs saying, 'Private yacht, no boarding'. A few of the old dockside cranes have been kept. They no longer have a function beyond the decorative; instead they have been absorbed into the heritage industry as background wallpaper. The gruelling manual work that used to keep the Island alive has been replaced by the sedentary trades of dealers, brokers and management consultants, whose tools are keyboards and mobile phones. They are the sort of people who once did the calculations that made the docks profitable, and who later made the calculations that closed them down.

All around the commercial buildings new blocks of luxury flats are being built. The developments have names like Dollar Bay, Discovery Dock, Cold Harbour, Ocean Wharf. A one-bedroom flat can cost £1 million. For four bedrooms you have to pay eight times as much. I wondered how many of the apartments would remain empty. Similar recent high-rise developments along the Thames have largely been gobbled up by the wealthy from China, Malaysia, Russia and the Gulf. Many of the flats are unoccupied, held as investments, or as bolt-holes in case of trouble at home.

~

A couple of weeks later I returned to the Isle of Dogs with the intention of walking right round the peninsula. I was curious to see whether there was any echo of the past still remaining beyond the private bounds of Canary Wharf.

Everywhere along the northwestern waterfront, behind high mesh fences, tall new buildings were going up, the central lift shafts rising first, the number of each floor stencilled in large numerals up the outside. But even the builders of this modern Babylon have to come to an accommodation with the elements. Down Limehouse Reach I crossed a footbridge over a creek thick with mud. The tide was out, and only a dribble of water trickled down the middle towards the Thames. Inland, the creek was held in check between concrete walls, disappearing into a low tunnel leading towards the skyscrapers of Canary Wharf. It was eerie to think of the sea creeping up over the mud at flood tide, infiltrating the foundations of the bastions of Mammon.

As I made my way south, a strand of gravel and broken bricks stretched along the shore, dotted with bits of water-worn glass and rusted lumps of iron whose former purpose could only be guessed at. Below the low-tide line there were rows of wooden piles, green with algae, the remains of long-abandoned quays.

Once this place had only been used by fishermen. Since medieval or even Roman times there had been a wall of mud and stones along the shore, protecting the Island from flooding. Along this wall were built the windmills that gave Millwall its name. But when steam power replaced wind power, and the docks were built, the fishermen and the millers were pushed out. The area became a hive of industry, with shipyards and ironworks, ropemakers and chain-makers, victuallers and brass foundries. There had long been a right of way along the top of the old wall, but in 1875 the factory owners, objecting to people walking through their yards, successfully petitioned for the right of way to be closed.

As I walked along this stretch of what is now the Thames Path, here permitted once more to keep company with the river, I passed a plaque commemorating the direct bomb hit at Bullivant's Wharf in 1941. Offshore, a police launch surged past a moored barge belonging to the Port of London Authority, loaded with some of the river's rubbish. Closer to the bank I watched a cormorant struggling with an eel. The bird managed to half swallow it, then the eel writhed back up the bird's long throat and splashed back into the water. The bird tried again, the eel wriggled out of its bill once more. The process was repeated until the bird eventually managed to keep the eel, seemingly still alive, down its gullet.

Out on the water, life and death are constrained only by predators, the police and the power of the river. Closer in, as it has always done, the tide claims and then gives up the foreshore, uncontrolled by human hand. But on land, walkers attempting to trace the river's bank frequently find themselves diverted by signs and barriers. Nowadays, it is not factory yards that people are prevented from entering. It is the private river frontage of luxury housing developments. The Thames Path may be a National Trail, but this does not mean that you can always walk alongside the Thames. 'Private', said the first sign, with a red arrow pointing inland. The next sign was more prolix:

Any use or occupation of this land or any part of it (including any roadway or pathway in front and behind this sign) by anyone as a right of way or by erecting or maintaining any sign is with our express permission only and that permission might be withdrawn at any time. We retain the right to possess and control every part of this land including by requiring any right of way or sign to be relocated at any time.

The notice was signed 'Express Newspapers'. And so it went on: 'No unauthorized access', 'Private property', 'No trespassing', 'Strictly no loitering, strictly no fishing', 'Private keep off grass', 'Secured by Protect Solutions'.

Sometimes the diversions took me into corners of the past. There are still some nineteenth-century terraces left on the Isle of Dogs, untouched by the bombs or the demolition squads. On Westferry Road I came across St Paul's, a small brick basilica in a highly decorated neo-Romanesque or possibly neo-Byzantine style. A faded plaque tells you in Gothic letters that it was affiliated to the Presbyterian Church in England. It is now used as a drama centre, but it was originally built to cater to the spiritual needs of the many Scots who came to the Isle of Dogs in the 1850s to work in the booming Millwall shipyards. It was the owner of one of these yards, John Scott Russell, himself a Scot, who laid the foundation stone of the church in 1859. Just the previous year, his yard, Messrs J. Scott Russell & Co., had launched what was then the world's biggest ship, the iron-hulled *Great Eastern*, designed by Isambard Kingdom Brunel. With a length of 692 feet, and a gross tonnage of 18,915, the *Great Eastern* was four times larger than anything else afloat. It was designed to take 4,000 passengers from England to Australia without refuelling, using a combination of steam-powered paddles and screws, assisted by sail. Close by the shore at Millwall you can still see the wooden piles and cross-pieces of the slipway from where the *Great Eastern* was launched. Such was the ship's great length that it had to enter the water sideways, rather than stern first, as was usual. The first few launch attempts failed, but on 31 January 1858, aided by an unusually high tide, the ship finally slipped into the water. Although during construction it had always been known as the *Great Eastern*, it was launched with the name *Leviathan*, after the great sea monster of the Old Testament.

This might have been hubris, on a par with César Pelli describing his Canary Wharf Tower as 'a door to the infinite'. Certainly, the ship seemed to attract misfortune. Both J. Scott Russell & Co. and the firm that commissioned the ship, the Eastern Steam Navigation Company, went bust. On the ship's maiden voyage, in July 1858, the ship, now once more christened the *Great Eastern*, suffered an almighty explosion as it sailed down the Channel. Below decks five stokers were killed by scalding steam. In just over a year Brunel himself was dead, and the ship's captain was drowned the year after that.

The *Great Eastern* never did take any passengers to Australia. The demand simply wasn't there. Instead, it spent a few years on the North America run, and later lay submarine telegraph cables. Misfortune continued to dog the former *Leviathan*, which ended its career as a mobile advertising hoarding for Lewis's department stores, until it was sold for scrap in 1888. As the ship was taken apart, two skeletons were discovered in the space within the double hull. The bones belonged to a shipwright and his apprentice who had inadvertently been sealed in as the ship was built all those years before on the Isle of Dogs.

The shipyards have long gone from the Isle of Dogs, even longer than the docks. One of the last of the yards to go was Thames Ironworks in the northeast corner of the Island (or just beyond it; definitions are hazy here), adjacent to the mouth of the River Lea. In 1898 the yard witnessed one of the worst peacetime tragedies ever to occur on the River Thames, during the launch of HMS *Albion*. When the battleship entered the water it created a huge wave that swamped a gangway carrying many spectators. Thirty-eight people, mostly women and children, drowned. Thames Ironworks closed in 1912.

Today, you rarely see a big ship on the Thames above the Estuary. What you do see are the speedy Thames Clippers taking commuters and tourists up and down the river. And there are police launches and pleasure craft. There are even still some commercial barges. But no big ships are built here, and no cargo is landed in the abandoned docks. There is a new trend, though: the cruise ship. The day I walked round the Isle of Dogs, Greenwich Reach was half-filled with an all-white nine-deck monster moored to a number of huge yellow buoys. This was MV *Viking Sky*, a brand-new vessel flying the Norwegian flag and part blocking the view of Greenwich's World Heritage Site – Wren's Royal Naval Hospital, Inigo Jones's Queen's House, the Royal Observatory, Greenwich Park.

But that was on the posh side of the Thames. Back on my side, in South Millwall, I was once more directed away from the river, past pubs called the Ship, the Lord Nelson, and the Ferry House, set amongst stretches of social housing that had survived the yuppie incursions to the north and along the shore. Down one of these more ordinary streets an over-muscled, shaven-headed man in white t-shirt and shorts, with tattooed arms and legs, walked his dog. I wondered (but didn't dare ask) whether he was a supporter of Millwall FC, which abandoned the Island in 1910 for a ground south of the river. Millwall supporters subsequently acquired a reputation for random and merciless violence, chanting 'No one likes us, no one likes us, no one likes us, we don't care.' The man with the dog was evidence that the white working class had not entirely abandoned the Isle of Dogs for the new towns of Essex. Although I noted that the dog the man was walking was not conforming to stereotype: it was not a Staff or an Alsatian or a mastiff. It was something small and fluffy. Perhaps a Pomeranian.

I rejoined the river at Island Gardens, at the southernmost point of the peninsula. Here my view of the glories of Greenwich – including the four-chimneyed, black-clad block of Greenwich Power Station, built before the First World War to power London's trams and Tube – was unpolluted by cruise ships. It was the view from here that is depicted in Canaletto's painting *Greenwich Hospital from the North Bank of the Thames*, although Canaletto himself (like T. S. Eliot) may never have actually set foot on the Isle of Dogs.

Not shown in Canaletto's painting is the post that in his time stood out in the river here. This was one of the places on the Thames where those executed for piracy were gibbeted. There is an iron gibbet cage displayed in the Museum of Docklands, and Dave Matthews told me all about gibbeting. Before being hanged, the convicted person would be visited in his cell to be fitted for his cage. In the eighteenth century this item cost a hefty £25, paid for, in the case of pirates, by the Admiralty (a cheaper alternative was to hang the body in chains). After execution, the corpse would first be tied upright to a post on the shore to be washed three times by the tide, in the name of the Father, the Son and the Holy Ghost. Thus purified, the corpse would be inserted into its close-fitting body-shaped cage, which was then suspended from the gibbet out in the river – far enough out so that the dead man's relatives couldn't rescue the body. The intention of the cage was to keep the displayed body intact for as long as possible, so that it would act as a warning to any passing mariners who might be considering a career in piracy. The head was covered in a sack to keep the birds off, and the rest of the body coated in tar to preserve it. The bolder of the local urchins used to swim out to the gibbet post in order to 'rattle the cage'. The aim was to keep rattling until the bones fell out.

Turning up the eastern shore of the Island, along Blackwall Reach, I came to a wide inlet. It was bordered on both sides by old wharves. Along its back, broad steps led down to a shingle beach. It seemed to be the place where the local Bangladeshi community came to enjoy the seaside. On the shore, two teenage boys in white shirts and dark suit trousers threw stones into the river, towards a trio of swans and the O2 Arena on the far side. Five young women, perhaps their sisters, all in hijabs and dressed from head to toe in various shades of pink, remained standing on the steps, each one attending to her mobile phone. To one side, a portly, bearded middle-aged father led his young son down to the water's edge, took a selfie and returned back up the shore.

Further on, towards the end of my circuit, I walked down the side of Millwall Inner Dock to catch the DLR off the Island, conscious of the sign that had just told me 'Pedestrian right of way round dock-side only'. Along the quay there was a succession of little wooden jetties extending out into the water. On every jetty sat a male mallard, each claiming his place. Perhaps this was the Isle of Ducks after all.

Further on there was a small, tilted wooden raft, covered in a mess of old nylon rope, plastic bags, and bits of stick and reed. In the midst of this, brooding the season's second clutch of eggs, sat a great crested grebe, splendid in its rust-coloured ruff and dark ear tufts. It seemed oblivious to the human world that had grown up around it, and which had almost wiped out its species in Britain in the nineteenth century when the bird was hunted almost to extinction for its head plumes. The grebe looked round warily, gave me a sharp glance, then tucked its head back under its wing. It knew it was safe from harm on its little island.

~

Later that summer, exploring further down the Thames, I walked eastward along the south shore from the old munitions factory at Woolwich. It was a hot August day, with a white sun hanging in a sky the grey of gun metal. Every now and again I would turn round and look back up the river, as if some sinister presence lurked behind me. Along the horizon, across the silver-splinted water, I could make out the skyline of the City and the towers of Canary Wharf, twin hubs of a worldwide web of finance and control. And I thought of *Heart of Darkness*, Joseph Conrad's oblique account of the colonial atrocities carried out over a hundred years ago in a faraway place then known as the Congo Free State. The tale is told by a mariner called Marlow aboard a yawl waiting for the turn of the tide down the Thames off Gravesend, across the river from Tilbury.

Before Marlow begins his story, Conrad sets the scene, looking back up the Thames, as I was looking now: 'And farther west on the upper reaches the place of the monstrous town was still marked ominously on the sky, a brooding gloom in sunshine, a lurid glare under the stars. "And this also," said Marlow suddenly, "has been one of the dark places of the earth."'

Conrad's novella points the finger away from Africa, long depicted as 'the dark continent', and towards the source of the evil, up the river towards London, the great imperial capital, its power and its pretence of civilisation built on the chain and the whip, its people sacrificed in the Blood Holes of its docks and yards.

When, many hours later, Marlow has finished his tale of deluded idealism twisted into exploitation and atrocity, his listeners have been so absorbed that they have missed the first of the ebb that was to take them out to sea. 'The offing was barred by a black bank of clouds, and the tranquil waterway leading to the uttermost ends of the earth flowed sombre under an overcast sky – seemed to lead into the heart of an immense darkness.'

Between the Heavens
and the Devil's Slide

Lundy

Lundy: Southwesterly 5 to 7, occasionally gale 8 later. Slight or moderate, becoming moderate or rough. Occasional rain. Moderate or good, occasionally poor.

– Extract from the Shipping Forecast, 2017

Fog and rain . . . Deluging showers . . . Windy, rainy, foggy chill . . . Wind and rain hideous . . . Days of storm and flood with breaks of windy sunshine.

– Extracts from the diary of the Reverend Hudson Heaven, owner of Lundy, September and October 1891

On the evening of 29 May 1906, HMS *Montagu*, a pre-dreadnought battleship, was conducting trials off the coast of southwest England in the new technology of wireless telegraphy. Although warships could now cross the Atlantic in a matter of days, and fire shells far over the horizon at unseen targets, communication between ships, or between ship and shore, had long been restricted to the limits of human vision. Messages to be sent were coded into strings of flags, which were then hoisted aloft – a method unchanged since Nelson had sent his famous signal before

Trafalgar a century before. Even in good visibility, and with the aid of a telescope or binoculars, such signals could only be read at relatively short distances. At night, signal lamps could flash messages in Morse code, but the range of such communications was limited by the fact that even from a crow's nest high on a mast, the horizon is no more than a dozen miles away. Beyond that, only telltale trails of smoke would suggest the presence over the horizon of another ship, whether friend or foe.

There was, however, another medium of communication that was unlimited by night, or fog, or the curvature of the earth. Since 1896 the British government had shown increasing interest in the experiments in radio transmission conducted by the Italian engineer Guglielmo Marconi. The Royal Navy was now involved in extensive trials, and on the day in question HMS *Montagu* had been ascertaining at what distance signals could be sent to and received from the Scilly Isles, off the western tip of Cornwall. As with signal lamps, messages were in Morse code, but the code was sent not as flashes of light but as pulses of radio waves, which were then converted into sound waves by the receiving device. A radio operator would listen to these short and long beeps through headphones, trying to pull the pulses down from a din of background static. What neither the operator nor anyone else at the time realised, was that this background static is a distant echo of the event at the very beginning of time – the Big Bang. Even after 13.8 billion years, you can still hear this echo, this hiss of white noise, any time you retune your radio between stations. It is as insistent, as unending, as waves breaking on a shore.

The trials having been concluded, Captain Thomas Adair ordered his ship to return to its anchorage in the Bristol Channel. By now, day had turned to night and the sea was blanketed in a thick fog, obliterating all signals from the lighthouses along the coast. Radio communication would still have been possible, but there were then no shore-based radio beacons to show the way. Adair and his navigating officer, Lieutenant J. H. Dathan, had to rely instead on compass and chart, on rulers, dividers and calculation. It was not an unusual circumstance; it was one they were well-trained for. Their intention was to make for Hartland Point, on the coast of Devon.

Shortly before 2 a.m. *Montagu* came alongside a pilot cutter. The pilot gave Captain Adair accurate directions for Hartland Point. Adair informed the man that he must be mistaken. As the battleship drew away from the cutter, the pilot shouted that if *Montagu* were to continue on its present course, it would encounter Shutter Rock within ten minutes.

Shortly afterwards, the awful sound of steel striking on granite could be heard through the fog.

The great battleship had run aground.

Adair sent a party ashore to look for help. They struggled up steep, rocky slopes, then made their way by lantern light along the top of the cliffs. They came across no habitation, until they at last came to a lighthouse. The officer in command of the shore party asked the keeper to inform the Admiralty that His Majesty's Ship *Montagu* had struck rocks south of Hartland Point. The keeper told the officer this was not Hartland Point. The officer insisted it was. The keeper in turn insisted that he knew very well which lighthouse he was in charge of. This was North Light, he said, on the island of Lundy, more than twelve nautical miles to the northwest of Hartland Point.

~

Lundy squats like a giant molar in the jaws of the Bristol Channel, a lump of granite three miles long by half a mile wide waiting to crush any ship that passes. It also gives its name to a sea area in the Shipping Forecast, broadcast four times a day, to warn all nearby vessels of adverse weather conditions. Countless voyages have been brought to an abrupt end on Lundy's jagged shores, the bodies of many drowned sailors dashed upon its rocks. Between the 1790s and 1890s alone, Lloyd's Registry of Shipping lists 135 shipwrecks on or around the island. We know few details of these wrecks, beyond the date of the loss and the name of the master. There are some exceptions. On 20 February 1797 a three-masted schooner called *Jenny* foundered at a place on the west coast still known as Jenny's Cove. She was said to have been carrying a cargo of ivory and gold dust. The ivory was salvaged, but the leather bags of gold dust have never been found.

For centuries Bristol was one of Britain's busiest ports, and so crowded a waterway was the Bristol Channel that at any one time 300 ships would have been in sight of Lundy. So it is not surprising that the island took a heavy toll on shipping. As early as 1786 the ship-owners of Bristol offered to build a lighthouse on the island at their own expense. Foundations were laid the following year. But it was not until 1819 that Trinity House (the body in charge of lighthouses in England and Wales) built the 'Old Light', at a cost of £36,000, on Beacon Hill – at 469 feet, the highest point of the island.

It was not an unqualified success. Low cloud or fog often sits on the island's plateau even when the air is clear lower down, so there were many nights when the light of the Old Light was not visible out at sea. The rate of wrecking barely diminished. In order to improve matters, in 1863 a gun platform was built at a place on the craggy west coast now called Battery Point. The idea was not to repel invaders, but to provide passing ships with an audible warning of danger in foggy conditions. Every ten minutes, the operators would fire one or other of two obsolete 18 lb cannons (dating back to the reign of George I). But this was not a universally recognised signal, so all too often ships believed they were under attack, and fled back out to sea. The Old Light and Battery Point both became redundant when two new lighthouses, North Light and South Light, were built on lower eminences in 1897, at the northwestern and southeastern points of the island. However, neither of these prevented the *Montagu* from smashing into Shutter Rock nine years later.

Subsequent attempts to salvage the *Montagu* were unsuccessful, and the following year the wreck was sold for scrap, and gradually dismantled *in situ*. During the salvage process, steps were built down the steep grass slopes above the cliffs to make access easier. These are marked on the Ordnance Survey 1:25,000 map as Montagu Steps. From the foot of the steps, those involved in scrapping the battleship would make their way along a narrow suspension bridge extending high above the waves out to the wreck of the *Montagu*, slowly grinding itself apart on the sharp rocks offshore.

Half a century before, the author Charles Kingsley, who had visited Lundy, imagined a fictional shipwreck on Shutter Rock. Towards the end of *Westward Ho!*, his swashbuckling yarn of Elizabethan sea-dogs, the hero is pursuing a Spanish galleon up the Bristol Channel in stormy conditions – squalls, rain and lightning:

On their left hand, as they broached-to, the wall of granite sloped down from the clouds toward an isolated peak of rock, some two hundred feet in height. Then a hundred yards of roaring breaker upon a sunken shelf, across which the race of the tide poured like a cataract; then, amid a column of salt smoke, the Shutter, like a huge black fang, rose waiting for its prey; and between the Shutter and the land, the great galleon loomed dimly through the storm.

The Spaniard realises the danger too late:

The galleon gave a sudden jar, and stopped. Then one long heave and bound, as if to free herself. And then her bows lighted clean upon the Shutter.

An awful silence fell on every English soul. They heard not the roaring of wind and surge; they saw not the blinding flashes of the lightning; but they heard one long ear-piercing wail to every saint in heaven rise from five hundred human throats; they saw the mighty ship heel over from the wind, and sweep headlong down the cataract of the race, plunging her yards into the foam, and showing her whole black side even to her keel, till she rolled clean over, and vanished forever and ever.

Shutter Rock, Jenny's Cove, Montagu Steps – Lundy is as bounded by toponyms as it is by its cliffs. Although the island is barely eleven or twelve miles in circumference, its coast is densely charted with names, each no doubt with its own story, though most are shrouded in the fog of time: Tibbett's Point, Miller's Cake, Ladies Beach, Brazen Ward, Frenchman's Landing, Virgin's Spring, Long Roost, Dead Cow Point, Mermaid's Hole, the Devil's Slide – to cite but a few. The proliferation of names suggests a need to list and identify every last feature of this difficult coast, to make clear where a precious cargo has been washed ashore, or where the body of a sailor lies, limp and broken.

But before humans arrived with their passion for naming, before even the ravens or the seals or the thrift or the heather came to the island, the rocks of Lundy rose from the depths of the earth.

Most of Lundy is made of granite. Lundy granite looks and feels like the granite of the tors and sea cliffs of Cornwall, but dating methods based on the ratios of naturally occurring isotopes in the rocks show that the granite of Cornwall and the adjacent Scilly Isles is some 280 million years old, much older than Lundy granite, which dates from less than 60 million years ago.

Granite is an igneous rock, one that cools slowly and solidifies deep underground, which accounts for its coarse crystalline structure – in comparison to denser igneous rocks like basalt, which have cooled more quickly (sometimes as lava on the surface) – hence their very much finer crystals, not visible to the naked eye. In the case of Lundy, the granite crystallised from molten magma possibly as deep as six miles underground. Across the island, the rocks are cut through with a swarm of dykes, intrusions of vertical sheets of basaltic rocks such as dolerite into the joints of the older granite. The dykes were formed between 56 and 45 million years ago as the Atlantic Ocean began to open up, splitting North America from Europe. This tectonic shift resulted in widespread volcanic activity that was also responsible for the spectacular basalt columns of Fingal's Cave and the Giant's Causeway. Lundy itself may very well have been covered by a large volcano, and the dykes seen today are evidence of molten rock erupting from deep in the earth's crust and cooling quickly beneath the surface. The volcano itself has long been eroded away by sea and wind and rain, leaving the granite, and the dykes, exposed. Lundy has a very strong positive gravitational anomaly, suggesting that dense basaltic magma still lies deep below the surface.

In relation to the slow passage of geological time, human presence on Lundy is no more than the blink of an eye. It is not until the Mesolithic period (beginning around 8000 BC, after the end of the last ice age) that the first evidence of nomadic hunter-gatherers on the island appears, in the form of scatterings of tiny flint blades. As the long Stone Age gave way to the Bronze Age in the second millennium BC, evidence of permanent human settlement on the island is found in the form of the remains of ovoid stone huts, field systems, standing stones and burial cairns. As well as growing crops, these people would have tended flocks of sheep, harvested seabirds and their eggs, and fished. Nothing much survives from the Iron Age or the period of Roman occupation of Britain, but in the cemetery on Beacon Hill, at the foot of the Old Light, there are four early Christian memorial stones, dating from the fifth or sixth century AD. Today, these semi-cylindrical granite stones rise out of the long grass of the graveyard, leaning against the cemetery wall, their backs to the prevailing wind. You can just make out a few inscribed characters, though it would take an expert to tell you that you are looking at memories of men and women with names such as Optimus, Resteuta and Tigernus.

Lundy may have become a minor religious centre in the early Middle Ages, and some of the isolated ruins above the cliffs on the western side of the island are possibly the remains of hermits' cells. There is no record of Viking incursions here, although it was the Vikings who gave Lundy its name, derived from Old Norse *lundi*, 'puffin', and *ey*, 'island'. The island does get a mention in the *Orkneyinga Saga*, in which a Hebridean–Norse leader called Holdbodi Hundason is forced to flee from his enemies, 'and did not stop until he came to Lundy'. From the perspective of the Icelandic author of the *Orkneyinga Saga*, Lundy must have seemed near the edge of the known world.

Around 1150, nearly a century after the Norman Conquest of England, Lundy came into the possession of the de Mariscos, an Anglo-Norman family. Some ten years later Henry II granted Lundy to the Knights Templar, but the de Mariscos ignored the royal decree and dug their heels in, determined, it seems, to be kings on their own island. In 1238 William de Marisco was accused of involvement in a plot to kill Henry III, but fled to Lundy, where he embarked on a career of piracy and brigandage. This was brought to an end in 1242, when the king sent a small force to storm the island's cliffs and seize William and sixteen accomplices. William was taken in chains to London, where he was convicted of treason, then hanged, drawn and quartered – the usual fate of traitors.

Henry installed a royal garrison on Lundy, and ordered the sheriff of Devon to build a formidable castle. In 1281 another de Marisco, son of the traitor's cousin, persuaded Edward I that he was not only a loyal subject but had a rightful claim to the island. He duly took over Lundy and its new castle, which to this day (considerably renovated) still stands high on a cliff near the southern tip, and which is still known as Marisco Castle – although the de Mariscos lost possession of Lundy in 1321.

Lundy has a perhaps undeserved reputation as a pirates' den, but pirates certainly operated out of it long after William de Marisco met his unpleasant end. The growth of Bristol as a port in the early modern period saw a great increase in shipping in the Bristol Channel, and this shipping, carrying precious cargoes from many parts of the world, was generally obliged to sail within range of Lundy, making it an ideal base for piracy. The island itself was easily defended: as the historian William Camden wrote in his *Britannia*, it is 'so encircled with rocks and cliffs round about that there is no avenue unto it but in one or two places'.

Many enterprising individuals saw the opportunities that Lundy thus provided. In 1610 a certain Captain Salkeld seized the island with piracy in mind and proclaimed himself king – a repeating theme among those who find themselves in possession of small islands. He amassed six or seven ships, and forced his prisoners to build fortifications. Salkeld's reign was brief. Shortly afterwards, while at sea, he is said to have been killed and thrown overboard by a fellow pirate.

A few years later the reputation of Lundy as an ideal base for freebooting had apparently reached North Africa, home of the Barbary corsairs (i.e. Berber pirates) who had long been raiding the coasts of the western Mediterranean for slaves. They had more recently extended their reach as far as the English West Country and the southern coast of Ireland; even distant Iceland was not safe. In 1625 they may or may not have occupied the island for two weeks, and two years later there are reports that a force from the Republic of Salé (in modern-day Morocco), led by their grand admiral, a renegade Dutchman called Jan Janszoon, took possession of the island and hoisted the Ottoman flag. The corsairs – known to the English as the Sallee Rovers – are said to have intermittently used Lundy as a base for some years, and kept some of their captives here before sending them for sale at the slave markets of Algiers.

At this remove, it is not possible to sift rumour from reality: there was certainly much pirate activity in the seas around Lundy in the sixteenth and seventeenth centuries, conducted by French, Flemish, Spanish, Basque and English ships, as well as Barbary slavers, but there appears to be no firm evidence that Lundy was ever used as anything but a temporary refuge and a lookout; it was not much use for victualing crews (compared to what could be seized from other ships or from ports), and offered no markets for stolen goods or sheltered locations for repairing ships. The island's owner in the early seventeenth century, Sir Bevill Grenville, never mentions the place being taken over by pirates, and instead in his letters portrays Lundy as an earthly paradise: 'I have so many reasons to be in love with it, as I shall never call or woo any man to buy it.'

Throughout the eighteenth century, Britain was more or less constantly at war with the French. To pay for these wars, the government imposed heavy duties on a range of imported goods, such as spirits, tobacco and tea. This in turn led many in the country's coastal areas to identify a new business opportunity: smuggling. For this purpose, Lundy provided an ideal entrepôt: being outside the six-mile range of Revenue cutters, smuggled goods could be landed here, and then transhipped to secluded coves on the mainland in smaller boats. Even those who were on the face of it respectable pillars of society were involved in smuggling, which became known as 'the gentleman's trade'. One such was Thomas Benson, a merchant and ship-owner, sheriff of Devon and MP for Barnstaple, who at some point in the 1740s took out a lease on Lundy and proceeded to use it as a base for large-scale smuggling, chiefly of tobacco. He was sometimes caught out, but despite facing fines totalling £8,229 (which he did not pay), he went on to ask the Commissioners for Revenue to establish Lundy as a regular port, with a pier, so that smaller vessels could shelter there. No doubt speechless at the cheek of the man, the Commissioners turned him down.

Benson was not short of chutzpah. In his capacity as ship-owner, he made contracts with the government to transport convicts to Virginia and Maryland, but rather than taking them to the New World, he landed them on Lundy, where he used them as slave labour in various construction projects, such as the building of a wall across the island (today there are three: Quarter Wall, Halfway Wall and Threequarter Wall; it was probably the first of these that Benson started). In his defence, Benson argued, successfully, that the convicts 'were transported from England, no matter where it was so long as they were out of the kingdom'.

A gentleman who visited Benson on Lundy wrote that 'The island at this time was in no state of improvement, the houses miserably bad . . .' Little of the ground was cultivated, being 'overgrown with ferns and heath'. The only food to be had was rabbits and seabirds, and there were so many rats 'that they destroyed every night what was left of our repast during the day'.

In the end, Benson tried one scam too many. After an unsuccessful insurance fraud involving a barely seaworthy ship, he was obliged to flee to Portugal, where he died in 1771 or 1772.

~

After the Benson era, Lundy changed hands several times. Various owners began with ambitions of improvement, but generally ran out of money before their visions could be realised. In 1787 a visitor reported that the derelict castle was occupied by two of the four or so tenant families, where they lived 'with great appearances of poverty and nastiness'. For a while the local economy continued to be based on smuggling, although the harvesting of rabbits, and seabirds and their eggs, was also important. The flesh of the birds was not eaten, but boiled down to make lamp oil.

The owner of Lundy from 1802 until his death in 1818 was the eccentric Irishman Sir Vere Hunt, who bought it on a whim at auction, having been assured that the island was free from both taxes and tithes, and that it was subject to neither king nor parliament. He drew up his own laws, and even oversaw marriages and divorces. But, like many of his predecessors, he failed to make any money out of his purchase, his wife remarking that 'Lundy Island has been productive of nothing but vexation.' Sir Vere himself left an account of his stay on the island in January 1811, when he was trapped there by inclement weather, recording 'High and piercing easterly wind and snow' on one day, while another was 'dreadful wet' with 'high wind and great fog'. It was so cold he slept in his greatcoat, and had recourse to burning sea pinks for warmth. When he did manage to embark for the mainland, 'the sea rolling very much', he and his party 'were all very sick'. At least Sir Vere managed to escape the island. His tenants and workers had no such choice. In 1821 one of them, a labourer called George Davis, wrote to Sir Aubrey de Vere (who had inherited the island from his father): 'I trust in God that your Honour will now order me to be paid, that I and my poor distressed family will quit this place, as the wretched habitation we now inhabit is in hourly danger of falling in and killing the whole family . . .'

~

In 1836 Lundy was purchased by a Gloucestershire gentleman called William Hudson Heaven, made rich by the income from the plantations in Jamaica he had inherited in 1820. It was out of the considerable compensation he received from the British government on the abolition of slavery in 1833 that he found the £9,870 required to buy Lundy, which he wanted as a summer resort for his family. As the owner of slave plantations, Heaven was used to wielding near-unlimited power over his domains. He appears to have treated Lundy in a similar vein, asserting, as his predecessors had, that the island was free from tithes and taxes. He went further, refusing to acknowledge any authority from the mainland and forbidding any official to land on Lundy. So – with that heavy-handed wordplay so beloved of the Victorians – Lundy became known as 'the kingdom of Heaven'.

Like his predecessors, however, Heaven found that his newly acquired kingdom was nothing but a drain on his dwindling finances, his income having been much diminished following the emancipation of his slaves. Although he built a modest Georgian villa (Millcombe House) in a sheltered dell on the southeast corner of the island, above the landing place, in 1840 Heaven decided to put the island up for sale. The advertisement placed in the *Times* perpetuated the myth of the island-kingdom: 'The possessor of this island becomes an instant sovereign lord . . . with an independence arising from the total absence of restraint or control . . . The mansion combines within it all the accommodation a patriotic little monarch can desire . . .' However, there was no sale, and the island stayed in the Heaven family.

William Hudson Heaven liked to live up to his name. He was a lay reader in the Church of England, and conducted services in his villa when he was on the island. Presumably his tenants were expected to attend. And, although government officials were refused access to Lundy, men of the cloth were welcome. One such, the Reverend John Ashley, made a number of visits. On 29 May 1842 he recorded in his journal a trip to Lundy to hold a service:

> The light keeper, Phelps, with his wife, refused to attend because they would not enter a building with any other person on the island. Such is the animosity that exists between them and has existed for upwards of twenty years past. There are not two persons on the island that will speak to each other or cross to each other's side of the island.

It seems the kingdom of Heaven was far from a paradise of harmony. Amidst a small population on a small island resentments can clearly build and brew, and feuds fester. Only fifteen years earlier, in 1827, Sir Aubrey's bailiff left this account of one of the inhabitants, a certain Mrs Davis:

> No man ever got such an abuse as she gave me . . . She then took a knife to run me through. I then had to give her a clout and a kick in the rump, which caused her to run off . . . She is a woman of the worst character. She keeps her house a house of ill-fame both with the pilots and sailors. She is both a H. [harlot?] and a lierd [liar?] and can never rest but provoking some person on the island.

The Rev Ashley, like many another Victorian missionary, was not one to be put off lightly. In his journal for 3 October 1842 he describes a return visit to the apparently God-forsaken island, an occasion when his calling was sorely tested:

> We came to anchor on the west side of the island, as it was not possible to land on the east side [there must have been a strong easterly blowing], and Mr Gibson and myself climbed the cliffs – to the inexperienced, as I found, a formidable under-taking. When I had got about half way up, becoming very giddy, and the cliff both above and below me almost perpendicular, I clung to the rock and determined to attempt no further, but the recollection of the painful state in which I found the inhabitants of the island . . . spurred me on, and I succeeded in gaining the summit . . .

His determination was rewarded that evening: 'The bells rung for service, and how different from former occasions. The island-ers, who would not formerly speak to each other, or assemble under the same roof, came together and filled the hall [of the villa] . . .'

The new spirit is perhaps reflected in the island's toponymy: alongside the fearful rock features of the Devil's Slide, the Devil's Chimney and the Devil's Limekiln, there now appeared on the west coast the Stones of St John, St James, St Peter and St Mark.

It was many years before Lundy acquired its own church. It was William Heaven's son and heir, the Reverend Hudson Grosset Heaven, who was responsible for the building of St Helen's. Completed in 1896, the church is a heavy exercise in Gothic Revival whose square tower and hunched body dominate the skyline of the southern part of the island. The Reverend Heaven could have built a harbour with the money, but presumably put the islanders' spiritual welfare above their physical well-being. Two years after his death in 1916, the Heaven family sold the island. The new owners did not last long, and in 1925 Lundy came into the possession of Martin Coles Harman.

~

Harman, the son of a builder, had first visited the island on a day trip in 1903, when he was just eighteen. He vowed then that he would one day buy Lundy. By 1925 he had made enough money working in the City that he did just that. He loved the island for its seclusion, its ruggedness and its wildlife. But with the ego of his predecessors, he declared himself 'King of Lundy', and maintained the island was 'a little Kingdom in the British Empire, but out of England'.

Harman forbade his tenants from owning radios, dogs or guns, and rejected the authority of all mainland institutions, dismissing from the island the GPO postmaster and the entire staff of the coastguard. He also decreed that no income-tax returns should ever be submitted by the residents; one elderly spinster who did so was asked to leave. The only outside authorities that maintained a presence on the island were the Church Commissioners, who owned the cemetery and the church, and Trinity House, who continued to manage the lighthouses. Harman went so far as to print special Lundy stamps (an institution that persists to this day). But when in 1929 he began to issue his own currency – half-puffin and one-puffin coins, worth a halfpenny and a penny (0.416p and 0.208p) respectively, with his own head on the obverse – he was prosecuted under the Coinage Act of 1870. Brought to trial at Bideford, he was fined £5, and obliged to pay costs of 15 guineas. He lost his appeal before the High Court in 1931, the Lord Chief Justice concluding there was 'no doubt that the English common law . . . applies on the island', so rejecting Harman's claim that Lundy had a similar status to crown dependencies such as the Channel Islands and the Isle of Man, which are not considered part of the United Kingdom.

As if playing God in the Garden of Eden, Harman decided to populate his island kingdom with some of his favourite animals.

The ponies (a cross between Welsh mountain stallions and New Forest mares), the Soay sheep and the Sika deer are all still there, as are the feral goats. Some of Harman's other introductions were less successful: the pair of mute swans planted on the island's only significant stretch of fresh water, the small pond called Pondsbury, flew off straight away, their wings still unclipped; the red squirrels refused to breed and had to be fed by hand; the peacocks were banned from the grounds of Millcombe House because they were too noisy, and dispatched to a mainland zoo; and the unfortunate wallabies either drowned in the island's few wells or were blown off the cliffs. Harman's devotion to natural history cannot be doubted; it was he who in 1946 helped to found the still thriving Lundy Field Society, which initially concentrated on the study of the birdlife of the island, notable for its many exotic migratory visitors.

In the 1930s tourism became the mainstay of the island economy, with sometimes several hundred trippers landing each day in the peak holiday season, carried by steamer from South Wales and North Devon. This development was encouraged rather than merely tolerated by Harman; he wanted to share the particular pleasures of his island with as many people as possible. Wealthier visitors arrived by aeroplane from an airfield near Barnstaple. Advertisements for Lundy described it as 'Britain's peaceful Gulf Stream Riviera', and assured prospective visitors 'No passports required'.

The outbreak of war in September 1939 called time on the Gulf Stream Riviera. The visitors left, and the resident men joined the Home Guard, arming themselves with shotguns, small-bore rifles and pikes. The Old Light was requisitioned by the Royal Navy as a lookout. Trenches were dug across any stretch of flat land where the enemy might attempt to make a landing. But otherwise Lundy played no great part in the Second World War.

Harman died in 1954. His son, Albion, took over the island, but following his death in 1968 the island was sold to the National Trust, and is now managed by the Landmark Trust, which owns and runs the MS *Oldenburg*, which from April to October runs between Lundy and either Bideford or Ilfracombe three times a week.

~

I'd read all about Lundy's reputation as a ships' graveyard before I embarked for the island, so it was with some trepidation that I boarded the MS *Oldenburg* in Ilfracombe's natural harbour. I'd also been warned that the crossing, of more than two hours, was likely to be a rough one, so I'd popped a couple of cinnarizine tablets in the hope of avoiding seasickness. I'd heard too many stories of the decks becoming awash with vomit within minutes of leaving port.

The ship was packed with perhaps two hundred passengers, taking up every seat below and on deck. Most were day trippers, some – like myself and my friends Jan and Bob – would be staying in a cottage on the island for a few days.

For a while we sailed parallel to the rugged coast of north Devon, past the cliffs of Brandy Cove and Bull Point, and the wrecking rock of Morte Stone, before setting out for the open sea. Behind us, as we sailed west, headland receded behind headland under a sky growing dark and lively with clouds and bursts of sun. To the south, Hartland Point jutted into the Celtic Sea, while to the north we could just make out the Gower Peninsula of South Wales. The sea itself, barred with shades of jade and ultramarine, was quiet, the waves low, my tablets of cinnarizine unneeded. A sign requested that when feeling ill passengers should use the sickness bags provided rather than the ship's toilets. Another sign warned of the dangers of the norovirus (the so-called winter vomiting bug), and asked passengers to 'sanitise the toilet area before and after use'. I found out later that for several weeks in 2007 the whole of Lundy was closed to visitors because of an outbreak of the norovirus, which is highly contagious, particularly amongst small, closed communities. The toilets in our cottage on the island bore similar signs.

Small islands are particularly vulnerable to introduced species, whether viral, floral or faunal. Lundy is no exception. Despite supposedly tasting of 'triple-distilled essence of Brussels sprout', the endemic Lundy cabbage is a favourite food of the goats introduced by Martin Harman, and with it are threatened two dependent insects, also found nowhere else in the world: the Lundy cabbage flea beetle and the Lundy cabbage weevil. The mauve-flowered *Rhododendron ponticum*, native to the Balkans and southwest Asia, was planted in the grounds of Millcombe House by the Heavens, but escaped and spread along the east coast of the island, crowding out the native flora and fauna. Over the last twenty years it has largely been eliminated by the Landmark Trust, using a combination of spraying and slash-and-burn. Visitors are requested to report any sightings, as the plant is extremely hard to eliminate. For many centuries rats (who even have a small island near the landing place named after them) preyed on the seabirds, in particular the puffins who gave the island its name, and whose habit of nesting in burrows made them especially vulnerable. By the early twenty-first century, there were only about ten pairs breeding on the island where once there had been thousands. A concerted effort over the last few years has seen the elimination of the rats on Lundy, and puffin numbers are slowly recovering. In the past, though, it was human beings who were the puffins' worst enemy. In 1816 Lundy exported 379 lb of puffin feathers to Ilfracombe, which, with twenty-four puffins required to produce 1 lb of feathers, meant that over 9,000 puffins had been killed on the island that year alone. By the 1880s the demands of fashion for puffin feathers had increased dramatically, and there are accounts of mainland fishermen visiting Lundy at the end of the close season, tearing the wings off still living birds, and tossing the bodies into the sea. Several hundred puffins would be killed this way in a day. As has been the case in many other parts of the world, the most damaging incomer species has proved to be *Homo sapiens*.

~

Once the *Oldenburg* leaves the Devon coast behind, another craggy coastline comes into view. This is Lundy, looming long and low on the horizon, its level skyline broken only by St Helen's Church and the Old Light. At first the ship seems to be sailing for the middle of the east coast, but then, having skirted the shoals off Rat Island, at the last moment it swings south into Landing Bay and the jetty beneath Marisco Castle and South Light.

At the back of the jetty there's a low neck of rocks and a stony beach which at low tide links Rat Island with the main body of Lundy. Peering southward through the gap, I could just see the head of a seal bobbing above the water on the far side. Its body would be hanging vertically below the surface, an activity called 'bottling'. It was September when I landed, the start of the three-month season when grey seals – at more than six feet long, Britain's largest mammal – come ashore on the island to give birth.

From the jetty, a track climbs steeply up the hillside to Millcombe House nestling in its woody dell, and then on up to the treeless plateau where the main settlement is no more than a small hamlet, unnamed, but boasting a shop and the Marisco Tavern. It is also the location of Old House, a fine granite building dating from around 1775. Part of this was to be our home for the next few days.

This being a ferry day, the Tavern was busy with day-trippers anxious for lunch. Lifebelts hang along the inside walls, memorials of more recent wrecks. Each lifebelt bears the name and often the home port or country of its ship: *Carmine Filomena*, Genova; *Taxiarchis*, Greece; *Waverley*, Bristol; *Devonia*, Bristol; *Amstelstroom*, Amsterdam. There's also an old sign advertising pickled gulls' eggs, a shilling apiece, and old black-and-white photographs of HMS *Montagu* in its pre-wreck prime, belching black smoke from its two stacks and flaunting its 12-in. guns.

I asked one of the women working in the Tavern what it was like living on Lundy, especially in the winter. She was dark-haired and handsome, perhaps in her late thirties. She loved it, she said. There was a real community. They didn't miss the mainland much.

Everybody got on with everybody else – in marked contrast to the discord found by the Reverend Ashley on his visit in 1842. I encountered a different perspective from a dark-haired, bearded man working in the shop just up the track from the Tavern. The shop sells everything from books to bacon, from t-shirts to tea bags. I asked him if he felt isolated in the winters on the island. 'I'm not here for the people,' he said, with a trace of a Welsh accent. 'This isn't the place for you if you don't like solitude.' He turned away, then said, 'It's lovely in winter.' I paid for my groceries and left.

On another occasion, I fell into conversation with a young woman serving behind the bar at the Tavern. She was only on Lundy in the summers, she told me, working as a conservation volunteer with the puffins and the peregrines. At one time, before she was born, both her parents had worked on the island, her mother as a chef, her father in conservation. 'My mum says I should have *Made in Lundy* tattooed on my foot,' she laughed. They'd always come here for holidays when she was a kid, after they'd left the island, and she fell in love with the place. She told me how she liked to snorkel off the east coast among the kelp forests and play with the seals. They'd come right up to her, she said, and pull her flippers. 'People get addicted to the place,' she said. 'It's a fantastic community.'

I got the same positive message on one of my walks from a young man with a wispy, reddish beard. He'd been on Lundy for three years as an assistant warden, he told me. He'd applied for lots of conservation jobs, including one at a place he'd never heard of: Lundy. He was surprised when he'd received a telephone call to attend an interview at Bideford, which he hadn't heard of either. Even though he didn't know where Lundy was, he'd got the job. He loves the island, he told me. He'd met his girlfriend here. There's one hitch, though. You have to leave the island if you have a kid, because there's no school, and no doctor. But that, I understood, was a problem for the future. In the meantime, he was living an idyll on what he described as a well-managed, unspoilt island.

Not everything, it seems, runs like clockwork on Lundy. One evening in the pub there was a small group of builders who were repairing the church. You couldn't miss the fact that the grim, grey granite of St Helen's was part-shrouded in scaffolding and plastic sheets. The men in the pub, still in their work clothes, were in something of a panic – or were joking that they were. It was already the third week in September, and there was to be a wedding in the church on the first of October. 'Down to him,' said a small man with a scar on his forehead, nodding at one of his companions. 'He does the slates.' The slater smiled.

'The bride and groom'll have to wear helmets,' he said. 'And high-vis vests.'

~

There was one thing above all others that I'd come to Lundy for, and it wasn't the church built by the Reverend Hudson Heaven.

It was the Devil's Slide.

The Devil's Slide is, to the climber, one of the most famous rock features in Britain, fit to stand alongside the Inaccessible Pinnacle of Sgurr Dearg on Skye, or the Old Man of Hoy. I'd climbed both the In Pin and the Old Man; the Devil's Slide had long sung me a similar siren song, ever since, aged fifteen, I'd seen photographs of it in *Mountain* magazine. That was back in 1970. I'd waited a long time.

Most of the rock features on the sea cliffs of Lundy are what you'd expect along a granite coast: deep zawns, cliff-top tors, towered ridges, jagged offshore pinnacles. The Devil's Slide, up the west coast beyond Threequarter Wall, is different. It's as if the Devil himself had taken a great cheese knife and pared off a vast slice of rock, leaving a smooth 45–50 degree slab of granite rearing 400 feet above the breaking waves of the Atlantic. There is nothing west of here for 3,000 miles, until you hit Newfoundland.

The Devil's Slide was first climbed in 1961 by Rear Admiral Keith Lawder, who after retiring from the navy had devoted himself to rock climbing, especially in the southwest of England.

Born in 1894, he had served as a midshipman at the Battle of Jutland, just ten years after HMS *Montagu* had steamed into Shutter Rock. There is a photograph of Lawder around the time of his first ascent showing a man in his late sixties with a plump, benign face, peering owlishly through round spectacles. Were it not for the scruffy, patched jacket, the deerstalker hat and the coil of rope draped over one shoulder, you might take him for a bank manager. The man must have needed some nerve to embark up the seemingly blank slab of the Devil's Slide. For a start, you can only reach the foot – a wave-swept platform only just above the sea – by a committing abseil. On the climb back up, with the equipment then available, run-outs would have been long and unprotected, and the belays tenuous. Today, even with better equipment making the route much safer, the climb is given the grade of 'hard severe'.

At one time, I would have looked forward to an ascent of the Devil's Slide as an undemanding day out at the seaside, a dance up easy-angled, sun-kissed granite. But two and a half years before my visit to Lundy I had had my accident, falling off Froggatt Edge. In addition to broken bones, I'd suffered serious brain injury, resulting in double vision: each eye presented a separate picture, each set at an angle pitched at 20 degrees to the other. The double vision gradually improved, to the point that it was (and still is) mostly only a problem when looking down.

Looking down. At your feet. That's what you have to do when you're climbing: scan the ripples of the rock and spot where there's a small fold or wrinkle that might support your toes. It wasn't too much of a problem at the indoor climbing wall, I'd found. There the holds are all colour coded. But on real rock, outside, there are no helpful colours. Everything is fifty shades of grey. Even a rough path or a boulder field can be problematic, requiring vigilance and a concentration on the fact that the data coming into the brain is often in conflict with itself. I still find it useful to shut one eye while descending uneven ground.

Uneven ground was exactly what I faced on the descent to the start of the Devil's Slide. The way down starts in a steep, grassy gully to one side of the Slide. Peering anxiously at where I hoped my feet were, I stepped from tussock to tussock, testing each foot placement carefully. Far below, the white-edged sea lapped at the base of the cliffs.

At the foot of the gully was a large lump of rock, round which Bob arranged a rope. The real cliff fell off from here. I told myself I was used to this sort of airy situation. I just needed to concentrate, follow through the procedures, make sure I was attached to the rope at all times, distil two conflicting visual images into one.

The final part of the descent took us by abseil a hundred feet or more down the slab of the Slide itself, to a huddle of blocks at its foot. These rocks jut out into the sea, forming one side of a narrow inlet bounded on the other side by St James's Stone, a jagged ridge extending out into the Atlantic, its granite ribs licked by the waves.

Once we were down, Bob led off up the Slide. I soon followed, happy to have the rope above me. My feet usually found somewhere that they could stick, as the angle of the rock is low and its friction high. But few of the holds, either for hand or foot, are positive, so balance was key. I've always liked slab climbing, as it requires more brain than brawn, and there is a delight in the way one adjusts the balance of one's limbs as one moves upward, requiring a constant awareness of one's body, eliciting pleasure in the knowledge that flesh can become a fluid thing, a process, always adjusting, extending, contracting, each part in communion with every other part.

Pitch followed pitch, the rock increasing slightly in angle. High above, the headwall of the slab reared over us like a breaking wave. The sky grew darker. The weather had started to turn. The morning had been clear, with a fresh breeze. Now the wind began to scud and bluster. Without large holds to grab onto, I had to anticipate each gust and steady myself against it, balanced delicately on the great slab. There were flurries of rain, too, enough for me to pull the hood of my jacket over my helmet, but not enough to soak the rock and make it slippy. The route made its way up shallow seams, around slight edges, tip-toeing on tiny protruding crystals, always demanding thought to find the surest way. Near the top, beneath the steep headwall, Bob led leftwards, then downwards into a corner, where he took a belay. Following, I had to balance along veins of quartz, then make a delicate downward traverse, the rope beneath me, to join Bob on his ledge. 'You can lead the rest,' Bob said. He'd been with me the day of my accident, and had refused to allow me to do any leading so far. He must have been confident that I was climbing safely again. The final, short pitch was entirely different from the sweep of bare slab below. It was a sort of blocky chimney, steep and rounded, but with cracks wide enough to jam your fists and feet into. This was as well, as now the rain and the wind were beginning to start up in earnest.

The rain and the wind continued to gather strength as we walked back along the track that cuts south down the back of the island. The shallow puddles that here and there covered the track shimmered with wavelets. In the distance, two miles away across the moorland, we could make out the Old Light, shrouded in a caul of drizzle and mist. With the wind in our faces, and my sack heavy with the rope, the Old Light seemed to get no nearer. I was glad I had my sticks to steady myself against the gusts, and then to drive myself onward.

We reached Old House at last, wet and tired, but content. I'd fulfilled a dream of nearly half a century.

Then my phone rang. It was my wife Sally. She had sad news. Time had run out for our dear old whippet Bertie, who'd been with us for nearly fifteen years. We knew he had kidney failure, was barely eating, was losing weight. Now he was stumbling, confused that his body was not working for him. The vet was to come to our house the following afternoon.

I told Bob and Jan. I knew they'd understand, they'd lost their old retriever not so long before. Bob also told me that when he'd been climbing on Lundy twenty years earlier, he'd had the news that his father had suddenly died. Jan, who had been at home, managed to phone through to the Marisco Tavern, and someone from there had found Bob in the small campsite. He told me he'd had to leave his companions and walk off on his own for a while.

~

My heart was heavy through my remaining time on Lundy. I tried to navigate a way round my imminent loss, scouring the corners and coasts of the island for comfort, from South Light to North Light. Both of these working lighthouses are now automated, their keepers long gone, their doors and windows bolted shut with plates of steel. Not far from South Light, on Shutter Point, I nervously skirted the Devil's Limekiln, a huge hole in the ground on the southwestern edge of the plateau. The hole drops like a mineshaft or a sudden death for a couple of hundred feet or more, down to a bouldery floor. There was once a horizontal passage from the bottom leading outward to the sea, but this has now been blocked by rock fall. Before that, in winter storms, with a southwesterly wind and the weight of the Atlantic behind them, waves would force themselves through the passage and jets of water shoot up the Limekiln, the spray spattering the grass and heather at the top. It's said that if you were to pick up Shutter Rock, just offshore, and turn it upside down, it would fit like a stopper into the Devil's Limekiln. With my uncertain vision I didn't dare go close enough to the edge of the hole to peer down. The Landmark Trust has made the decision not to erect fences or warning signs; those visiting Lundy must rely on their own judgement. My judgement was that the rock around the rim of the hole was rotten, so I kept well clear.

I visited the Old Light, climbed its 147 spiral steps. At the top, where the lantern would have been, there is a small platform equipped with two striped deckchairs for visitors to rest and enjoy the 360-degree view. At the foot of the Old Light is the island's ancient graveyard, located, not next to the relatively recent church, but around the stumps of granite that mark the ancient graves of Optimus, Resteuta and Tigernus. There are also Victorian gravestones, grey-patched or hairy-green with lichens. One of these recalls Samuel Jarman, 'who fell over the cliff on Lundy Island December 24th 1869'. This would not have been the season for collecting eggs, so who knows how he met this misfortune; perhaps a gust of wind took him over the edge; perhaps life on the island proved unendurable. Then there are more recent burials, including that of Martin Coles Harman and his wife Amy Ruth. Their graves are marked by small granite boulders, uncut and uninscribed, but each bearing a plaque. Much of the area bounded by the drystone cemetery wall is overgrown with long, lush, windblown grass curving over shapes that might be burials, or which might simply be rocks.

On the day the vet was to visit our house, we walked up the east coast, following the narrow path that skirts across the slopes above the cliffs, while the wind and the rain combed the plateau above. Around the tors, ravens wheeled about, guarding their nests, while out at sea the occasional gannet, sharp white wings tipped with black, patrolled above the waves. Down below on the rocks uncovered by a falling tide two grey seals basked, exposing their bellies to the air, even though there was no sun. One scratched its head with a flipper. In the water a third seal bobbed its head above the surface, then dived to play beneath the waves. Out of the corner of my eye a peregrine made a diagonal slash across the air close to the cliff face, then was gone, leaving the arc of its flight behind. The sky was emptied of other birds, as if a squeegee had been drawn across an insect-spattered windscreen. Pigeons and other prey won't fly if they sense a falcon somewhere above, ready to fold itself into a stoop so fast that a bird hit by it would be dead before it knew what had happened.

The morning of the day our ferry was to depart, I explored on my own. I started out in drizzle and low cloud, to match my mood of loss. As I walked down past Millcombe House, the sky began to clear. On a small nearby hill a Union Jack rattled officiously from a tall flagpole, blown by a fresh southerly breeze. As the surface of the sea below began to sparkle and the bracken turned rust-gold on the slopes, a seal offshore slapped its flipper on the surface and then rolled over. Flags and other such declarations of dominion seemed pitiful human conceits in the face of the slow unwinding of time. I sat down in the sun by some stunted oaks, and focused my senses. All I could smell was the sweet coconut scent of gorse in flower, all I could hear was distant birdsong and the wind troubling the leaves of the oaks above me. Further away there was the endless shushing of the waves, sounds without meaning, without beginning or end, the white noise of the Atlantic.

~

As we queued that afternoon to board the *Oldenburg* I overheard a conversation between a middle-aged man and his father. They were looking towards a large lump of something on the shingle beyond the jetty.

'Wossat?' asked the father.

'Seal,' said the son.

'Alive or dead?' asked the father.

The seal – a large brown bull – gave a loud snort.

'Alive,' said the son.

Then the wind blew his hat off. He chased after it, but it was too late. It tumbled off the edge of the jetty into the sea.

I smiled as I boarded the *Oldenburg*, whatever the emptiness waiting for me at home; smiled as we sailed past Mouse Island, its rocks crowded with seals. Some of the cows would have been ready to give birth. The white-furred pups grow quickly, gaining some 4 lb in weight per day on their mothers' fat-rich milk, putting on a thick layer of blubber in readiness for the time, four or five weeks after birth, when they will abandon the shore and slip into the water, ready for a life at sea. Awkward on land, in their own element grey seals are sleek and agile, feeding on sand eels and mackerel, pollack and wrasse, playing in the forests of kelp, diving as deep as a hundred fathoms into the twilight far beneath the surface.

As we watched the island slowly recede under a mother-of-pearl sky, I felt I'd left a part of me behind. Perhaps it was just my absence that I'd left behind, as others had marked their absences: a footprint in the sand, a headstone in a cemetery, a crumpled letter, a flicker in someone's memory – like Samuel Jarman, the man who'd fallen off one of Lundy's cliffs on Christmas Eve 1869, leaving only his grave to remember him by; like George Davis, fifty years before, begging to be paid the money owed him for his labour, so that he could quit the island and the wretched habitation he was forced to live in; like his namesake (and possibly wife) Mrs Davis, who was reduced to selling herself to passing sailors, her bitterness sowing discord around the island; like Benson's slaves, forced to break rocks to build a wall with no purpose; like all the forgotten, nameless generations before, eking out a living scrabbling for eggs, for seabirds, for fish; those sparse generations whose unrecorded bones are scattered round the island, slowly dissolving in its acid soil.

It was going to be a long, sad way back to a home diminished by loss. In the growing swell, the *Oldenburg* heeled from side to side. It was two days past the equinox, and the autumn storms would not be far behind. As I sat by the port rail I watched the starboard beam heave 20 degrees above the horizon, then swing back down. I was glad I'd taken two more cinnarizine tablets before setting sail.

Behind us, to one side of our wake stretching back towards Lundy, a single gannet turned itself into a missile and harpooned into the water, leaving no trace but a splash.

The tide drains out of one of the creeks along the shore of Canvey Island (Ian Crofton)

The proprietor of the café at Paradise Beach, Canvey Island (Ian Crofton)

One of the painters of the sea-wall mural commemorating the North Sea Flood of 1953, which killed fifty-eight Canvey Islanders (Ian Crofton)

Right. After the last party on Eel Pie Island, near the top of the tidal Thames (Ian Crofton)

Below. *La vie bohème* on Eel Pie Island. Most of the houses hide behind high fences and locked gates. (Eden Breitz/Alamy)

Top. Looking from the Isle of Sheppey across the Medway towards the Isle of Grain (Ian Crofton)

Left. A pub by the former naval dockyard in Sheerness (Ian Crofton)

Below. Second World War fortifications guarding Sheerness docks (Ian Crofton)

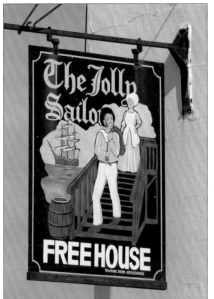

Aerial view of Wallasea Island. The eastern end is being restored to wetland by the RSPB. (David Wootton/Alamy)

Looking up the Thames towards the Canary Wharf development on the Isle of Dogs (Ian Crofton)

Walking the dog in Millwall, Isle of Dogs (Ian Crofton)

The slave-owner who founded the West India Docks.
The statue was removed in 2020. (Sentinel 3001/Alamy)

The last of the shipyards on the Isle of Dogs. It closed in 1912 (Ian Crofton)

'Lundy squats like a giant molar in the jaws of the Bristol Channel' (Paul White Aerial Views/Alamy)

The author climbing the Devil's Slide on Lundy (Ian Crofton collection)

The landing place on Lundy. Rat Island is to the left of the jetty (Ian Crofton)

Top. The tidal channel between Mersea Island and the mainland (Ian Crofton)

Above. Beach huts at West Mersea, the populated end of the island (Ian Crofton)

Left. Colchester natives are mostly reared in beds around Mersea Island (Ian Crofton)

Above. The castle on Piel
Island, built by the wealthy
monks of Furness Abbey
(Ian Crofton)

Right. Beacon on the
shore of Morecambe Bay,
with Piel Island in the
background (Ian Crofton)

Below. The Lindisfarne
causeway, with the refuge
for those caught out by
the tide (Helen
Hotson/Shutterstock)

The ruins of the twelfth-century priory on Lindisfarne (Antonia Gros/Shutterstock)

The fishermen on Lindisfarne use upturned boats as sheds (Ian Crofton)

Lindisfarne Castle, a Tudor fort turned into a medieval fantasy by the architect Edward Lutyens (Philip Bird/Shutterstock)

Aerial view of Tresco, with Bryher beyond (Timothy Dry/Shutterstock)

Subtropical vegetation drapes the ruins of the medieval priory in Tresco Abbey Gardens (Ian Crofton)

The south shore of Tresco, with, in the distance, the Eastern Isles (centre) and St Mary's (right) (Ian Crofton)

The Troy Town Maze on St Agnes (Ian Crofton)

Chesil Beach, the only natural land link to the Isle of Portland, just visible in the distance (Last Refuge/Alamy)

The lighthouse on Portland Bill, guarding some of the most dangerous waters on the south coast (Adrian Baker/Shutterstock)

The Verne on the Isle of Portland – once a citadel, now a prison (Meibion/Alamy)

Hilbre Island, with the windfarms of Liverpool Bay lining the horizon (Ian Crofton)

Middle Eye (Little Hilbre Island), with the coast of Wales beyond (Ian Crofton)

The lighthouse on Longstone, from where Grace Darling set out on her dramatic rescue (ATG Images/Shutterstock)

The Farnes are home to tens of thousands of guillemots (Ian Crofton)

Looking up the east coast of the Isle of Wight, from the Landslip (Ian Crofton)

Coastal erosion along the Back of the Wight, near Blackgang Chine (Ian Crofton)

The Needles, with the Isle of Purbeck in the distance
(Ian Crofton)

Pirates and smugglers loom large on the Isle of Wight
(Ian Crofton)

Two of 'Palmerston's Follies', Spitbank Fort and Horse Sand Fort, artificial islands in the Solent (Ian Crofton)

Samphire and Oysters

Mersea Island

Poor Britons – there is some good in them after all – they produce an oyster.

> – Sallust (Gaius Sallustius Crispus), attributed (*c.* AD 50)

'Between the mouths of the Blackwater and the Colne, on the east coast of Essex, lies an extensive marshy tract veined and freckled in every part with water.' So begins *Mehalah: A Story of the Salt Marshes* (1880), one of the more favourably regarded novels of the Reverend Sabine Baring-Gould, best remembered as the author of 'Onward, Christian Soldiers'. He continues:

> At high tide the appearance is that of a vast surface of Sargasso weed floating on the sea, with rents and patches of shining water traversing and dappling it in all directions. The creeks, some of considerable length and breadth, extend many miles inland, and are arteries whence branches out a fibrous tissue of smaller channels, flushed with water twice in the twenty-four hours. At noontides, and especially at the equinoxes, the sea asserts its royalty over this vast region, and overflows the whole, leaving standing out of the flood only the long island of Mersea . . .

In 1871 Baring-Gould had taken up a new living as the rector of East Mersea, bringing with him his young wife Grace, the daughter of a Yorkshire mill hand, together with the first few of their fifteen children. When he'd met Grace he had been a thirty-year-old curate; she'd been just fourteen. It was thought the Church of England hierarchy had posted him to remote East Mersea to prevent him spreading his somewhat unconventional ideas among more civilised people.*

East Mersea is the scarcely populated half of Mersea Island, the half that faces the mouth of the River Colne where it meets the Blackwater Estuary. Baring-Gould found his few parishioners 'dull, reserved, shy and suspicious', and admitted 'I never managed to understand them, nor they to understand me.' Nevertheless, he stayed on Mersea for ten years, and was inspired by the island and its people to write the melodramatic *Mehalah*, published in 1880 and described by one critic as 'the Wuthering Heights of the Essex salt marshes'. Mehalah herself is the young and spirited daughter of a tenant farmer on the Ray, an islet that sits between Mersea and the mainland, 'a hill of gravel rising from the heart of the marshes, crowned with ancient thorn trees'. Mehalah, Baring-Gould tells us, 'might have been taken for a sailor boy', had not 'the evening sun lit her brown gipsy face, burnt in her large eyes, and made coppery lights in her dark hair'.

The original inspiration for Mehalah might have been a local girl called Mahala, who would have been thirteen when Baring-Gould arrived on the island. Mahala was the second daughter of William Baker, who ran the foot ferry from the Mersea Stone – the eastern tip of the island – across the mouth of the River Colne to Brightlingsea. Baring-Gould describes in his *Reminiscences*

---

* Baring-Gould was probably more eccentric than dangerously radical. He had the habit of teaching with a pet bat clinging to his shoulder, and his 1,240 publications range from lives of the saints to a study of lycanthropy. Baring-Gould was often confused by his numerous progeny, once asking at a children's party, 'And whose little girl are you?' 'I'm yours, Daddy,' the little girl wailed. His marriage to Grace lasted happily for forty-eight years, and gave George Bernard Shaw the inspiration for his play *Pygmalion*.

visiting Baker and Mahala's stepmother Anne on the old barge where they lived:

> We went often to Baker's boat and had tea with him. Whilst tea was brewing he would set a pail of shrimps before us, on the deck, and bid us fall to until the tea was ready in the cabin. Baker himself was a sober man, but his wife was often tipsy. When she returned late from Brightlingsea, overcome with liquor, Baker hauled up the ladder. He emptied a pail of water over her head, as she stood shouting for admittance, and left her to scold, swear and shiver till he considered her to be sufficiently sober to be admitted.

Mahala herself worked as a domestic servant in Brightlingsea, later married an Irish soldier, and died of consumption in 1893, at the age of thirty-five.

My own first visit to Mersea, more than a century after the death of Mahala, followed the route taken by Mahala's father's ferry, from Brightlingsea across the mouth of the River Colne to the Mersea Stone. Although a passenger ferry still runs along this route, I crossed the channel sailing a sturdy Wayfarer dinghy. It being high tide, we could beach our boat on the shingle above the mudbanks, and laze awhile in a strange flat landscape carpeted in low bushes. The bushes looked like untamed heather, but turned out to be something called shrubby seablite, a woody plant with fleshy, salt-resistant leaves. Here and there among the seablite grew brittle thin fingers of marsh samphire, interspersed with the mauve flowers of sea lavender, once thought effective against 'colic, dysentery, strangury, and fluxes of the blood'. But our idyll could not last long. If the ebb tide left our dinghy dry on the mud, it would be many hours before the flood returned to get us afloat again.

~

Today, as in the days of Mahala/Mehalah, there's only one way to get to Mersea Island if you're not going by boat, and that's via the Strood. The Strood is a causeway (or 'causy' as the locals call it) that carries the only road from the Essex mainland onto the island. This road becomes impassable (or at least inadvisable) for an hour or so either side of high tide, when the Strood becomes awash with seawater.

As sea levels rose after the end of the last ice age, it is possible that ancient people built a wooden trackway woven out of alder branches across the tidal marsh and mudflats that separate Mersea from the mainland. But the earliest evidence of a causeway here consists of rows of timber piles, dated to around AD 700, and hence Anglo-Saxon in origin.

There is no straightforward route to the Strood from the main A120 road heading east from the M11 to Colchester. Given that the latter was an important Roman town, and that some of the Roman inhabitants are believed to have holidayed on Mersea, it is surprising that there is not a dead-straight Roman road linking the two. Instead, I followed a maze of minor roads through deepest rural Essex, jinking this way and that round field boundaries, passing through villages with names such as Tolleshunt Knights and Great Wigborough. As the minutes to high tide on the Strood ticked down, I nervously thought of Chesterton's line, 'The rolling English drunkard made the rolling English road', and wondered whether I would make it in time. So it was with some relief that I saw at last the Strood stretch out before me, clear and dry, with cars streaming across to the island. It was a sunny Saturday, and everybody was heading for the beach.

As I drove over the causeway I could see that the sea was high either side, brown with wash-off from the recent heavy rains. A strong wind was blowing, and, although it was a bright July day, choppy waves were breaking white against the edges of the island's salt marsh. A sign on the road advised that on a certain Sunday in August traffic might be delayed by boats crossing the Strood. This is the climax of the annual Round-the-Island Race, in which sailing dinghies are manhandled across the causeway by teams of volunteers.

Once I reached the island itself, another sign – possibly unique in the annals of prohibition and threat – declared:

PRIVATE MARSH
KEEP OUT
SAMPHIRE PICKERS
WILL BE PROSECUTED

Marsh samphire has had something of a boom among foodies in recent years, either eaten raw or steamed for a few minutes. Tasting a bit like salty asparagus, it makes an ideal accompaniment to many seafoods, or on its own, served with butter and lemon. Its growing popularity – and price – is presumably the reason for the sign that greets you as you arrive on Mersea. Its alternative name, glasswort, alludes not to its brittleness and translucence when young but to the fact that it was formerly burnt to make soda ash, used in the manufacture of glass. The word 'samphire' itself comes from the French *herbe de Saint Pierre*, 'herb of St Peter', he being the patron saint of fishermen. Another, rarer, name for the plant is 'mermaid's kiss', presumably alluding to the taste of the sea it imparts. Marsh samphire is not to be confused with the unrelated rock samphire, which gets a mention in *King Lear*, when Edgar pretends to the blinded Gloucester that they are standing on top of a cliff near Dover:

> The crows and choughs that wing the midway air
> Show scarce so gross as beetles: half way down
> Hangs one that gathers samphire, dreadful trade!
> Methinks he seems no bigger than his head . . .

Rock samphire was once as popular a culinary item as its marsh namesake, having, according to Culpeper, a 'pleasant, hot and spicy taste', but its consumption has long fallen out of fashion. Perhaps the price was just too high. Robert Turner, writing in 1664 of rock samphire growing on the unstable cliffs of the Isle of Wight, wrote that 'it is incredibly dangerous to gather; yet many adventure it, though they buy their sauce with the price of their lives'.

I had once collected and eaten marsh samphire from the maze of muddy creeks near Blakeney in north Norfolk, where there was no threat of prosecution; so the day of my visit to Mersea, having already tasted the forbidden fruit, I was happy to pass by the 'private marsh', and leave its samphire unpicked. My first destination was East Mersea, in particular the crumbling low cliffs of compacted gravel that face out across Brightlingsea Reach, laid down 300,000 years ago when the Thames was a tributary of the Rhine. I'd read that coastal erosion here had revealed the fossils of hippopotamuses and straight-tusked elephants; I wondered whether I might also find rock samphire alongside the marsh samphire that grows in the flats by the nearby Mersea Stone.

I was to find neither fossils nor rock samphire. Instead, walking down the shorn, parched grass to the shore from the car park of Cudmore Grove Country Park alongside numerous families out for a day by the seaside, I encountered a pillbox. It was a typical Second World War pillbox – hexagonal, concrete, low to the ground, with gunslits on each side. But it was also a thing transformed. Where once it had stood as an ugly utilitarian necessity to resist invasion, now it might have been a piece of abstract sculpture, weathering into the landscape. Young children clambered onto its roof, crawled into its innards, peered out of its loopholes, while their parents took photographs.

This pillbox was originally part of the Eastern Command Line, one of several defensive lines constructed in the early days of the Second World War against possible German invasion. The Eastern Command Line ran from Mersea Island up the River Colne, through Colchester and northward into Suffolk. Mersea had long been of strategic importance: in the early eighteenth century Daniel Defoe wrote that

'Tis thought a thousand men well provided, might keep possession of it against a great force, whether by land or sea. On this account, and because if possessed by an enemy, it would shut up all the navigation and fishery on that side, the government formerly built a fort on the southeast point of it. And generally in case of Dutch war, there is a strong body of troops kept there to defend it.

There is little now to be seen bar earthen banks of the 'fort on the southeast point', a blockhouse built and extended in the sixteenth and seventeenth centuries. Most of the Second World War defences have proved similarly transient. When I walked westward from the pillbox along the shore beneath low cliffs of soft orange rock, I found many slabs of concrete and chunks of broken brickwork, all that remains of a 4.7-in. gun battery that once stood along the cliff top, but which has now fallen victim to coastal erosion. The Scots pines that now line the cliff top are suffering a similar fate: after a few hundred yards my way along the shore was blocked by a tangle of fallen trunks and branches. Far in the distance, across the Blackwater Estuary, stood the twin blocks of the now decommissioned Bradwell nuclear power station. In the other direction, east across Brightlingsea Reach, lines of windmills were spinning in the sunshine on the flats of St Osyth Marsh.

I retraced my steps. The cliffs ran out, and a path took me eastward along a sea wall, into the seablite-covered flats leading towards the Mersea Stone. Once I reached the easternmost point of the island, I turned back, this time following the shore. The breeze was stiff, the water choppy. Small, impatient waves broke frequently by my feet. My lurcher-whippet Zigzag, a rescue who had replaced Bertie, found a dead crab in the shingle. Pouncing on it with glee, he seized it in his jaws, then scampered off, crunching. Then he suddenly stopped. He had seen something. There was some *thing* rolling slowly at the edge of the sea, washed gently up and down by each wave as it seeped in, seeped out.

A few other walkers passed by on the shore. 'What do you think it is?' a young man asked me. His girlfriend had wrinkled up her nose, though there was no smell.

'Dead seal,' I suggested.

Zigzag was peering at the thing, cautiously stepping towards it. I called him off, then approached more closely. The body, about three or four feet long, lay on its side. The belly was white, the back dark grey. The tail had been torn off, the genital and anal area ripped into one deep, dark-blue gash. The skin on the chest and round the face had been sliced into hanging flaps, exposing areas of greyish-pink flesh. The eyes were gone. The body was more streamlined than that of a seal. There was a dorsal fin, and a mouth lined with small, even teeth.

It wasn't a seal. It was a porpoise.

I'd once seen a porpoise while sailing in these waters. It had rolled playfully up out of the water a few times, a sleek live thing.

Zigzag approached the body once again, cautiously sniffing. To him it was an unknown, alien object, some kind of animal no doubt, moving but not alive, severed from its element. He backed off, looked away. I'd never seen a dog so disconcerted by death. Or perhaps it was my own certainties that were shaken.

A pair of ringed plovers scurried hither and thither across the shingle, then flickered quickly away with the wind.

When I told the warden at the country park about the corpse I'd found, he said he already knew about it. They were waiting for the council to take the body away. He seemed embarrassed by the whole affair. 'Shouldn't a vet examine it to determine the cause of death?' I asked, thinking it might have fallen victim to some mysterious virus. I'd seen seals dead of canine distemper on the Suffolk coast a little to the north. 'Oh, we already know what killed it,' he said. 'Hit by a propeller.' The estuary of the Blackwater is full of pleasure craft. But porpoises are a rarity.

~

There were a few more things I hoped to see on Mersea. The map had marked a vineyard, and that was definitely on my list. I turned down a side road, and into a car park. There were a few rows of vines, and beyond them a big marquee. That's where the tastings will be, I said to myself. But when I got out of the car, I could hear music, and round me milled men in suits and women in long dresses. The marquee was for a wedding, not for wine. I wandered up to a group of buildings. The only sign welcomed me to Maria's Vintage Tea Room. It wasn't quite what I had in mind, but I went in. A teenage lad was at the till. I asked him whether they sold wines from the vineyard. 'Behind you,' he said, indicating a few shelves with bottles.

'Can you tell me about the wines?' I asked. 'What grapes do you use?'

'No idea. There's a leaflet, I think.' There was. There was even a wine called 'Mehala', made from Ortega grapes. For some reason I instead chose 'Island Dry', which uses Reichensteiner. There were no tastings on offer, so I bought a bottle to take home. Later, once it was chilled, I tried it. Very sharp on the first sip, then deliciously dry and fruity. It would have gone well with oysters.

Oysters. I'd heard that Mersea is famous for them. They're known as Colchester natives. I think of them as Colchester consolations, salty bivalve bringers of delight, slipping past the palate, then down the gullet. West Mersea is the place to go. All along the coast there, past Cobmarsh Island, along the Strood Channel towards the Ray, the map marks 'oyster pits', shallow pools filled with sea water where oysters were formerly placed to turn them 'green about the gills' – that colour once being valued by epicures (the green is in fact caused by an alga rather than by putrefaction). Oysters have been harvested around Mersea since Roman times. Indeed, the Roman historian Sallust remarked 'Poor Britons – there is some good in them after all – they produce an oyster.'

West Mersea is now a sprawl of brick-built suburban streets, cheered up by rows and rows of pastel-painted beach huts along the southern shore. Past these, turning up the channel called Besom Fleet, the island begins to roll up its sleeves. This is where the oyster fishery is based. It's also where I found the West Mersea Oyster Bar. The place was packed. I asked whether I could buy half a dozen natives. The young man said he couldn't open them for me to take home. 'High-risk food, oysters,' he explained. 'But I can open them if you eat them now. You can sit outside.' And so I sat outside, and relished the fleeting pleasure.

~

I couldn't leave the island without visiting Baring-Gould's parish church in East Mersea, dedicated to St Edmund King and Martyr. Edmund was a king of East Anglia before England was united, and, according to legend, met his death in 869 at the hands of the pagan Vikings who had taken him captive. The Danish commander Ivar the Boneless and his brother Ubba demanded Edmund renounce his Christianity, and when he declined they had him beaten, shot with arrows and finally decapitated. His head was thrown into thick forest, and his family and followers were only able to find it with the guidance of a spectral wolf, which showed them the way by crying (in Latin) *Hic, hic, hic* ('Here, here, here'). The church itself dates back only to the twelfth century, and I could find little reference to the shady Edmund, although in a niche there is a modern wooden carving of the saintly martyred king with a wolf sitting loyally at his feet, having its muzzle fondled. Nor could I find any reference to Baring-Gould, although no doubt the congregation still sing his hymns.

At the gate to the churchyard I'd noticed a small sign saying 'At this location there are Commonwealth War Graves'. I wandered around, expecting to see rows of white crosses and wreaths of red poppies. There were none. Old stones bore mottoes such as 'Long affliction patiently borne' and 'Peace at the last'. One stone remembered a little girl, 'interred in Quetta Cemetery, India'. She had lived less than a year. Then I found a stone in loving memory of George Cudmore, 'who died from wounds received in France, April 9th 1917, aged 34 years'. Beneath George, the stone also remembered his brother Nathan, 'who died in France from wounds received in action, March 21st 1917, aged 29 years'. The name Cudmore rang a bell. Then I remembered that I had just visited the Cudmore Grove Country Park, down by the East Mersea shore. It was there I had found the dead porpoise, another creature torn from the vigour of its young life. Who knows what state the bodies of George and Nathan were in when they died. War graves don't record such things. Perhaps as they crossed over they heard, or imagined they heard, someone singing to them: 'Onward, Christian soldiers, marching as to war . . .' Except for them there was no 'as'. They'd marched to war, and the Cross of Jesus had not saved them.

Out in the Blackwater a speedboat buzzed, its propeller slicing the waves. I thought of the two brothers, dying messily, slowly, of their wounds. I hoped the porpoise had died a quicker, cleaner death.

Quicksands, Cocklers
and Killer Whales

The Isles of Furness

Sinking water . . . Many, many sinking water . . . Sinking water, sinking water.

— 999 call on 5 February 2004 from one of the twenty-one Chinese cockle-pickers drowned that night in Morecambe Bay

Arnside is a pretty village tucked snugly up the northeastern armpit of Morecambe Bay. The last time I'd visited the place had been fifty years before, when, with the youngest of my three older sisters, I'd stayed for a week with two great-aunts in the bungalow they shared in their retirement. They'd never married; the men of their age had been decimated on the Western Front, leaving a generation of maiden aunts.

I have a memory of sober hats, dark coats and sensible shoes, of quiet, respectable Lancashire voices, and quiet, respectable games of cards, all confined within beige and brown interiors. It was an odd environment for two children on the cusp of adolescence. I was twelve, and still pliable, and hence not fully aware of the stifling atmosphere. My sister was nearly fifteen, and seethed in silence. I'm not sure the aunts knew what to make of us. We certainly did not know what to make of them. We were tolerated, even loved as family in a dutiful Edwardian way, but perhaps not entirely approved of. This was the Easter of 1967, the Easter just before the Summer of Love. Muhammad Ali had recently refused to be drafted to fight in Vietnam. The great-aunts did not approve. The young men they'd known, including two of their brothers, had gone to their deaths without question or complaint.

Morecambe Bay became a lesson. There were rules that must be adhered to. Despite the beauties of the sands and the water, we were warned, the Bay was a place of great danger, of quicksands where seemingly solid ground dissolved in a moment, sucking the unwary under. And if you weren't caught in a quicksand, we were told, you might find your feet stuck in heavy mud, while the incoming tide raced towards you at the speed of a galloping horse. My sister and I looked out across the Bay with a mixture of fear and breathless anticipation.

The dangers have not diminished over the succeeding half century. All along the promenade that lines the shore there are warning signs:

EXTREME DANGER
BEWARE
FAST RISING TIDES
QUICKSANDS
HIDDEN CHANNELS
IN AN EMERGENCY DIAL 999 – COASTGUARD
SIREN WARNS OF INCOMING TIDE

At low tide the sands of Morecambe Bay stretch for miles and miles, split by the channels of a number of rivers – the Kent, the Bela, the Eea, the Leven. Other channels are so transient and unpredictable that they remain unnamed. Some of the sands have names – Greenodd, Milnthorpe, Warton, Ulverston, Rooseback; some are banks – Mort Bank; others are wharfs – Cartmel Wharf, Yeoman Wharf. Here and there are scatterings of seemingly solid gravel, mixed with boulders and cobbles: Cowp Scar, Sea Wood Scar, Church Scar, Elbow Scar, Point of Comfort Scar, Barren

Point Scar. All these scars are submerged at high tide, so never quite achieve the status of islands, places of safety at all points of the tide. One of the more infamous of the scars is not a scar at all, but a skear: Priest Skear, off Hest Bank.

On the night of 5 February 2004 the emergency services received a 999 call from a mobile phone. The voice at the other end could only repeat, in broken English, 'Sinking water . . . Many, many sinking water . . . Sinking water, sinking water.' It was known that a team of Chinese cockle-pickers were out in the Bay. Local pickers had seen them as the sun went down, and tried to warn them, tapping their watches to indicate time was running out before the tide raced in. In the pitch black of that February night the RNLI deployed its Arnside hovercraft, a vehicle well-suited to the untrustworthy terrain, where in a moment water becomes land, and land becomes water.

As the crew of the hovercraft began their search of Warton Sands they encountered what one later described as 'a sea of bodies'. Many of the crew hadn't seen corpses in the water before. Over the next twenty-two hours they were to recover the bodies of twenty-one men and women, aged between eighteen and forty-five. They were scattered over a wide area; it seems many had tried to swim for the shore, but had been overcome by hypothermia and drowned.

The coastguard helicopter, meanwhile, used thermal imaging to detect any sign of life in the cold dark of the Bay. There were no such signs to be found – until, on Priest Skear, the infrared camera picked up a man frantically waving his arms as the tide rose round him. The helicopter shone a searchlight, and a life-boat picked him up. The man's name was Li Hua. He had just failed to save his friend, whom he called Brother Wen, from drowning.

Li Hua, the only survivor, became a key witness in the investigations and trial that followed. The cocklers had all been illegally trafficked into the country by Chinese Triads to work at rates much lower than those paid to local cocklers. Among those convicted, the Chinese gangmaster was sentenced to twelve years for manslaughter. Two men from Merseyside who had bought the cockles were cleared of helping to break immigration laws.

The scars of Morecambe Bay only provide temporary refuge to those caught out by the tide and the fickleness of the forever shifting channels. But there are more permanent sanctuaries in the Bay, collectively known as the Islands or Isles of Furness. Furness (or at least that part of it called Lower Furness) is the peninsula that bounds Morecambe Bay to the west. It was, until 1974, an exclave of Lancashire, and the name means, in Old Norse, 'headland by the island shaped like a backside'.

Among the few Isles of Furness, Chapel Island and Sheep Island are tiny and unpopulated. Roa Island and Foulney Island are now mere spits of the mainland. That leaves three populated islands: the small Piel Island, accessible only by boat; Barrow Island, once separated from the town of Barrow-in-Furness by a wide channel, now narrowed into a succession of docks; and the long strip of the Isle of Walney, England's eighth largest island, joined to Barrow by a single bridge. Some scholars believe Walney (which possibly might be 'the island shaped like a backside' alluded to in the name of Furness) may mean 'island of the killer whale': the suffix *ey* is certainly Old Norse for 'island'; the *Waln-* element may be *vogn*, the Old Norse word for a killer whale. The Vikings who named the island *Vogn Ey* may have been thinking of Walney's shape; it seems they could not tell their arse from their orca. The cetacean similarity has been noted by more recent commentators: in *Lost Lancashire* (1922), Arthur Evans describes Walney as 'shaped like a gigantic, stranded whale'. It is one of those ironies of history that over the last century or more, on Barrow Island just across the channel from Walney, the Royal Navy has built most of its fleet of submarines, including those that stalk beneath the world's oceans carrying the UK's nuclear deterrent. The shape and capability of these submarines echo, on a gargantuan scale, those of the killer whale.

It was to reach the Islands of Furness that I found myself staying a night in Arnside, at Ye Olde Fighting Cocks, a pub I am sure that neither of my great-aunts ever set a foot inside. That evening I walked along the shore of the Kent Estuary, taking in the views north to the Lakeland Fells – views that had clearly made no impression on me fifty years before, because I had no memory of them. Every now and again, two-carriage passenger trains trundled backwards and forwards over the Kent Viaduct that stretches across this neck of Morecambe Bay. This is the Cumbrian Coast Line, which links Arnside and the rest of England to the isolated towns and villages of western Cumbria – Ulverston, Barrow, Millom, Sellafield, Whitehaven, Workington, Maryport, Wigton and so to Carlisle.

As the light faded across the falling water, I saw two locomotives pulling a pair of flat-bed trucks across the viaduct. On each of the trucks there was a much smaller, yellow, box-like container. I wondered whether the freight – clearly a very heavy kind of freight – came from Sellafield, the fenced-off, heavily guarded plant on the remote western coast of Cumbria where nuclear fuels are reprocessed. It was here that the UK first recovered fissionable plutonium for its atomic weapons programme. The box-like containers I saw may have contained some kind of nuclear waste. The official government website for Sellafield says that

> highly active liquid waste . . . is 'vitrified', meaning it is dried to a powder and mixed with glass at a temperature of around 1,200 degrees Celsius. The molten mixture is then poured into stainless steel containers and allowed to solidify. We store the waste in a specially engineered store, pending its final disposal in the UK, or return it to its country of origin.

In the morning the tide was out, and the glassy ribbon of the River Kent cut through the sands and under the railway viaduct making its slow way to the sea. Two men were parked up in their van on the esplanade, unpacking fishing rods. They told me they were after flounder. Apparently they grow quite large in the Bay. Far to the north, I could just make out the peaks of the Langdale Pikes. I had ahead of me a long drive round the indented northern shores of Morecambe Bay. It is a strange, winding route. Sometimes it feels like you are in the midst of the Lakeland Fells. At others, you are skirting the flat expanse of the mud-strewn sea. At one point I was surprised to see, on top of a craggy hill above the road, a lighthouse. Once I was through Ulverston, my way stuck to the shore. Far to the south, across the Bay, I could make out the twin blocks of Heysham nuclear power station. In the foreground, waders clustered in the shallows where an unnamed stream wound between the sands. Lines of intertwined tyre tracks stretched across the wet flats towards the horizon. I assumed they were made by the tractors of cockle-pickers, although all I could see far out in the Bay was a line of twisted, four-legged navigation towers, leading ships towards the port of Barrow.

With the tide out, Piel did not at first look like an island. I could see the sandstone castle, and the white-washed Ship Inn, and the row of pebble-dashed pilots' cottages. It seemed I was only separated from them by a stretch of salt marsh and an inlet of mud. Beyond the castle lay the length of Walney Island, which from the mainland seemed part of the same landmass as Piel – which it is at low tide, although it would be foolhardy to attempt to cross on foot.

It was only when I drove across the causeway to the pimple of built-up land called Roa Island, and walked from the small car park along the road to the top of the jetty, that I could see that Piel Island, even at low tide, was cut clear of the mainland by the deep expanse of Piel Channel. Piel Channel leads from the open sea past Pickle Scar, Pike Stones Bed, Long Rein Point and Cove o' Kend all the way up to the docks at Barrow. It is deep and wide enough for ocean-going ships, and (although this information will be classified) even the largest of unseen submarines.

On the steps at the top of the jetty sat an elderly, fit-looking couple with a dark brindled greyhound called Zoe. I asked the woman about Foulney Island, the spit next to Roa Island. 'If you're going to go there, you'd better be quick. The tide'll cut you off in an hour or so.' She told me that only the day before a dog had had to be rescued from Foulney. And the local lifeboat was always being called out by people trapped there by the tide. I decided to give it a miss. There was nothing marked on the map but mud, sand and stones, and beyond that Blackamoor Ridge, Farhill Scar and Foulney Hole. I asked the woman if she'd ever been trapped by the tide. 'Once,' she said. 'When I was a kid. I got stuck in the mud under Walney Bridge. Me dad rescued me. But not me wellies. They're still there.'

'I grew up in Vickerstown,' she continued, referring to the model estate built for their workers on Walney Island from the 1890s by the shipbuilders Vickers, who by then dominated industry in Barrow. She'd worked for thirty years as a buyer of parts for Vickers, then their successors BAE. She said she didn't feel like she was ever on an island.

I asked about the ferry to Piel. 'The ferryman should be here soon,' she said. 'People will begin to line up at the end of the jetty.' In a few minutes a van drew up behind us. 'Bathbrite', it said on the outside. 'Restoration of all baths.' It turned out that the driver – the proprietor of Bathbrite – was also the ferryman. 'Not enough money in the ferry,' he told me. 'So I travel all over the country fixing baths.' It turned out he owned the only three machines that can restore antique enamel bathtubs, so he is in demand far and wide. 'Better get down there quick,' he said, pointing along the jetty, where a couple of families had congregated. 'They're queuing up.' And with that he hoisted a small rowing dinghy not much bigger than a bathtub over his head and trotted down the steps. I followed him along the jetty. His head and body were covered by the dinghy. All I could see of him were his wellington boots. Surely he wasn't going to row us all across the channel to the island in *that*? The distance was at least a third of a mile, and the water looked deep and the tide fast flowing. I began to worry, but reassured myself that we were just beside the RNLI station.

Once he reached the end of the jetty, the ferryman tipped the dinghy into the water, jumped in and rowed off at speed. In a while he returned at the helm of a twenty-foot motor launch, which he'd fetched from its mooring. He handed us in one by one, and then we curved across the channel, the engine puttering effortlessly. In a matter of minutes we came to the jetty on Piel Island.

I was the first to disembark. As I walked up the jetty a Union Jack flew from a flagpole at the top of the shore, snapping in the wind. I could see two figures sitting at the tables arranged outside the Ship Inn. I approached them. 'The King and Queen of Piel, I presume?' I asked.

'Indeed we are,' said the King.

'I am pleased to make your majesties' acquaintance,' I said. 'Might it be possible to purchase a pie and a pint?'

By tradition, the landlord of the Ship Inn is known as the King of Piel. Although the inn only dates back to the late eighteenth century, the island's monarchy traces its lineage back to the time of the Tudor king Henry VII and his young rival Lambert Simnel, the pretender to the throne of England who purported to be Edward, seventeenth Earl of Warwick, the Yorkist claimant. In June 1487 Simnel and a force of 2,000 German (or Flemish) mercenaries landed on Piel Island from Dublin. From there they made their way south to take on the army of Henry VII at the Battle of Stoke. The rebels were defeated. Rather than executing his rival, Henry (a master of spin) put Simnel to work in the royal kitchens as a turnspit – an unpleasant and lowly job. The Kings of Piel still work in hospitality, and the institution (which may only date back to the eighteenth or nineteenth century) has an air of mockery about it. When a new King is anointed, he sits in an old chair wearing a battered helmet and clutching a bent sword, while buckets of beer are poured over his head. The current incumbent is Steve Chattaway. His Queen is called Sheila.

There is a castle on Piel – a vast, ruined castle – but it long predates even the time of Lambert Simnel. It was built by the powerful monks of Furness Abbey, once among the richest foundations – and biggest landowners – in England. They built the castle to protect their profitable trade with Ireland and the Isle of Man – not only from pirates and freebooting raiders, but also from the officers of the crown who might attempt to impose duties on imported goods. In the end, the crown seized the castle – and all the lands of Furness Abbey – on the Dissolution of the Monasteries under Henry VIII.

It doesn't take long to walk round Piel. I wondered at the sheer scale of the castle, its multi-coloured sandstone blocks eroded into twisting patterns by wind and sea and rain. The crumbling masonry remains as a memorial to a long-gone power in the land. The only other structure on the island, apart from the castle and the inn, is the row of pilots' cottages. There were a few abandoned boats lying about, one planted up with marigolds and petunias, and a broken-down Land Rover with columbine growing out of its radiator grill. Everything looked more than a little abandoned. Under a warm sun and a hazy sky, martins flew low over the long, late-August grasses. Oystercatchers peeped, and in the distance a curlew called.

I returned to the Ship Inn for my lunch. There wasn't a pie on the menu, so I ordered the falafel burger. I snatched a few exchanges with the King and Queen, anxious not to distract them from their hostly duties to the boatloads of trippers the ferryman had by now delivered to their shores. Their majesties told me that only two other people – a retired couple – live on the island permanently, in one of the pilots' cottages. The other three cottages in the row are part-time holiday homes. The King and Queen stay on the island all through the winter, when the ferry doesn't run. They don't have a boat themselves. I said they must be special kinds of people to survive the isolation. 'We're living the life,' said Queen Sheila.

'We love it,' said King Steve.

On my return voyage to Roa Island there was a younger man at the helm. He told me his name was Jordan Cleasby. He was, he said, the son of John Cleasby, who'd taken me over in the morning. Jordan said his father had run the ferry for thirty-six years. Before that his grandfather Colin had plied the same trade. Jordan and his parents and his siblings all live within a few streets of each other in Barrow. It's clear they are a close family. Jordan works in Barrow docks, on the tugboats that take cargo ships in and out of the lock gates. His shifts often last twelve hours, but he still finds time to help his father with the ferry.

Today, the only lock gates linking Barrow docks to Walney Channel – then Piel Channel, and ultimately the open Irish Sea – are situated at the southwest of what is still called Barrow Island. There was once another set of lock gates at the northwest end. Originally the channel between Barrow Island and the mainland was wide enough to justify the place being called an island. But when the docks were built in the nineteenth century – to replace the less sheltered port at Roa Island – much land was reclaimed from the sea, and the channel narrowed. Along with the docks, the nineteenth century saw Barrow expand into a major industrial centre, with extensive shipyards and (by 1876) the largest steelworks in the world, using locally mined iron ore. One contemporary, stunned by a view of the steelworks by night, wrote of 'vast pyramids, from whose red summits rise huge cones of clinging flames and domes of smoke'. The last remnant of the steelworks closed in 1983. Locals still remember the great slag heap by its side glowing in the dark.

Barrow had been a tiny village, without an available workforce, so the new industries brought in thousands of migrants from Ireland and Scotland. From almost the beginning, Barrow specialised in submarines. In 1886 it launched the *Abdül Hamid*, commissioned for the navy of the Ottoman Empire. The *Abdül Hamid* became the first submarine to fire a live torpedo underwater. In the later twentieth century Barrow built the submarines that carried the UK's Polaris, then Trident, nuclear missiles. The northwestern end of Barrow Docks is now blocked off by BAE's Devonshire Dock Hall – the giant shed used to hide what goes on inside from the prying cameras of hostile spy satellites. The shed was built in the 1980s on part of the Devonshire Dock that was filled in with 2.4 million tons of sand pumped from Roosecote Sands, to the south of the docks. In these parts, land and sea are interchangeable.

Just beyond the Devonshire Dock Hall the A590 crosses the Jubilee Bridge to Walney Island. The bridge was only built in 1908. Despite the fact that so many of Barrow's shipyard workers now lived in Vickerstown, the construction of a bridge had long been resisted by the Furness Railway Company, which ran the ferry to the island. It was not keen on competition. The only alternative to the ferry before the bridge was built was a line of stepping stones across Walney Channel, the reliability of which can be judged from its name: the Widow's Crossing. Two rights of way are still marked by the Ordnance Survey from the mainland across Crook Scar and the mussel beds to its north to Walney Island, but a right of way does not necessarily mean a safe passage. Keeping on the right side of caution, I drove across Jubilee Bridge. It was just another busy urban road.

Walney Island when you first encounter it is built up. Once across the bridge you are in Vickerstown, a rather strait-laced place lined with Edwardian terraces, a petrol station and a branch of Tesco Express. I could see no sign of a pub. The 2011 census records that 96.9 per cent of the population of Barrow (which for such purposes includes Walney) are White British. My great-aunts would not have felt out of place here. I did. And so I took the small road south, past Tummer Hill Marsh with its blistered wooden sign:

PUBLIC ARNING
Q ICKSA D AREA
KE P LEAR

I drove on along the top of Biggar Dyke, down Mawflat Lane, past Robin Whins Point, Whylock Marsh and the caravan site at Far South End to the dunes of South Walney Nature Reserve. The road ends here. There is a marked trail through the dunes, and there are frequent requests to keep off the shore for the sake of the birdlife. The dunes are well-covered with bracken and marram grass, and the occasional wind-blasted hawthorn. Here and there I saw small bluish-purple patches of viper's bugloss. It was too late in the year for the rare, indigenous Walney geranium (*Geranium sanguineum* var. *striatum*), a pinker, stripier variant of bloody cranesbill. Beyond the dunes, half hidden by them, stood a hexagonal white lighthouse. Among the dunes lay a pattern of brackish lagoons. On the side of one there was an unmanned oyster farm. A cormorant stalked along a strip of grass in between the lagoons, while sandpipers fluttered low over the water, flustered and peeping.

Walney being a narrow island, in no time I came to its western shore. Along the horizon, far out over the glittering sea, stood line after line of windmills. At the mouth of Morecambe Bay nature is now generating safer, softer energy than at Sellafield or Heysham. Walney itself is said to be the windiest lowland location in England. On a flat grass bank above the dune line that marks the shore I found splinters of wood, piles of shattered timber, evidence of the fury of the wind and the sea – the same power that makes the windmills turn. Facing the other way, I could make out BAE's Big Shed. Beyond it, in the distance, crouched the massive rounded hump of Black Coombe, the Lake District's most westerly fell, looming as a permanent reminder of humanity's impermanent folly.

It was getting late, and I could not linger. That night I had to be in Langdale, where the sea is nothing but a memory. As the late-August sun tipped towards the horizon, I began my journey north, along the Duddon Estuary, past Coniston Water and into the heart of Lakeland.

Twice Isle and Twice
Continent in One Day

Lindisfarne

On the coast of Northumberland over-against the river Lindi,
we see Lindisfarne . . . which (as Bede says) is twice isle, and
twice continent, in one day; being encompassed with water at
every flow, and dry at every ebb; whereupon, he calls it very
aptly a semi-isle.

– William Camden, *Britannia* (1610 translation)

In the warm July of 2013 I was walking the border between
Scotland and England when I came across a path cutting at right-
angles to my route. A waymark sign said this was St Cuthbert's
Way. I, like the border, was heading north, following the watershed
ridge of the Cheviot Hills. The sun shone, and golden grasses
waved gently in the late afternoon breeze.

I'd passed the Schil and Black Hag, and found the sign just
beyond Tuppie's Grave. To the west, St Cuthbert's Way sloped
down into Scotland and the small town of Yetholm, eventually
reaching Melrose Abbey, where Cuthbert – originally a shepherd
boy from the Lammermuir Hills – had become prior around AD
662. To the east, St Cuthbert's Way descended into England to
meet the College Burn at Hethpool. From there the Way heads for
Lindisfarne, where Cuthbert became prior some three years later.

In Cuthbert's day there was no border here. This land, from the Lothians to the Tyne and beyond was all part of the Kingdom of Northumbria.

At one point I'd considered walking to Lindisfarne along St Cuthbert's Way, as a kind of pilgrimage. But it's a path many others have followed, and the Way itself is a recent creation. As it turned out, chance, fate and the constraints of the calendar conspired to take me a different route.

After visiting the Isles of Furness, I'd spent a damp weekend climbing in Langdale. It was early September, and the long hot summer of 2018 was over. By the Monday, the weekend's drizzle had turned to downpour. I looked at the map. Although from the perspective of London, the Lake District is in the far northwest of England, on its east side England nudges even further north, taking the Cheviot Hills and then the River Tweed as its northern boundary. It looked a long way from Langdale to Lindisfarne, which lies off the Northumberland shore only a few miles south of Berwick. Eschewing the obvious routes along major roads, I decided on an oblique traverse of the Pennines, from Cumbria northeast into Northumberland. The fact that it was pouring with rain and the cloud was right down would only add to the sense of a quest.

Not far beyond Penrith the road began to climb up the western flanks of the Pennines. On the lower slopes the hay had been cut but not gathered; the torrential rain was ruining the harvest. I hardly passed another car as I rose into the gloom shrouding the hills. At one point through the murk I saw a sign to Croglin, a remote hamlet at the foot of Scarrowmanwick Fell. The name rang a bell, and then I recalled the story of the Bat of Croglin. In 1875 a young woman called Emily (or Amelia) Cranswell, while staying at Croglin Low Hall, endured a horrific experience. One night a manlike creature in a black cloak broke through her bedroom window, pinned her to her bed, and proceeded to bite her about the face and throat. The creature, Emily (or Amelia) reported, smelt of decay. On a subsequent night, Emily's (or

Amelia's) brothers stood guard by her room, and when the creature appeared they pursued it to the churchyard, where it dived into a vault. When the vault was opened, the brothers found a decomposing corpse – with fresh blood on its mouth. The creature was dispatched with a stake through its heart. The story is in all probability the creation of the Victorian writer Augustus Hare, and has been subject to a number of variants. But as I drove into the darkening clouds beneath the hill still called Fiend's Fell, I was not inclined to scoff. Fiend's Fell lies only a few miles north of Cross Fell, the loftiest point in the Pennines, where in the fifth century St Paulinus had raised a cross to exorcise the demons that then inhabited the bleak, high moors.

Today, the upper slopes of this western escarpment of the Pennines are covered with the grey stumps of cleared conifer plantations, interspersed with scatterings of more recently planted native species. Above the tree line, all I could make out was thick bracken fading up into the mist. Along both verges there were rows of striped poles with red reflectors, to mark the course of the road when deep under snow.

Eventually I made it to the top. A sign told me I was on Hartside Summit, at 1,904 feet. There was little to be seen in the mist, bar the ruin of Hartside Café, which had burnt down a few months earlier. Hartside Summit marks the watershed of this part of England. To the west, Loo Gill becomes Raven Beck, a tributary of the Eden, which flows into the Solway Firth and thence the Irish Sea. To the east, Rowgill Burn feeds into Black Burn, which at Alston joins the River Nent to form the South Tyne, which in turn joins the North Tyne before entering the North Sea at Newcastle.

From Alston I followed the road northeast over the pass by Willyshaw Rigg that marks the Northumberland border, and so descended into remote Allendale. Beside the road, the purple flowers of rosebay willowherb had mostly turned to fluffy seed. The upper hillsides were bare, bar reeds and thistles growing thick among the withering grass. Clumps of mist drifted across the dark face of a distant forestry plantation.

Allendale, like Croglin, has a story of a fearsome, spectral creature. In 1904, several farmers reported that they had found numbers of their sheep dead on the fells. They'd been mauled, and bore the marks of large teeth. One local swore he'd seen the silhouette of a giant wolf stalking the skyline. The man was disbelieved – until a certain Captain Bain of Shotley Bridge near Consett let it be known that a wolf he owned had escaped, and that a reward was available for its capture, dead or alive. Hunters tramped the fells with guns, but found no trace of a wolf. Was it all a hoax? It wasn't. To the west, along the railway line between Settle and Carlisle, a body of a wolf was found. It had been hit by a train. The wolf was stuffed and put on show at the railway company's HQ in Derby. A contemporary song celebrated the tale, ending:

> If thoo dissent believe me
> Or think Ah's tellin a lee,
> Just ax for the wolf at Derby,
> It's there for a' to see.

I thought of these stories as I drove on my pilgrimage towards the Holy Island of Lindisfarne. 'Hobgoblin, nor foul fiend,' I said to myself, 'Can daunt my spirit . . .' As a would-be pilgrim I was, I knew, a bit of a fraud. But I had a mission, and a destination.

So I drove on through the mist and the drenching rain, crossing the narrow stone bridge over the River Allen in its wooded gorge. Signs warned of rock fall. Further on I crossed the Tyne, then the line of Hadrian's Wall, then the dead-straight rollercoaster of Dere Street, the Roman road from York to the Firth of Forth, now mostly taken by the A68.

The departure of the Roman legions at the beginning of the fifth century AD left a vacuum in Britain – both temporal and spiritual. The Celtic Romano-Britons left behind were by now Christian, but their power and beliefs were largely eclipsed by the Germanic pagans – Angles, Saxons, Jutes and others – who swept in from across the North Sea. It was these pagans, whom Celts and Romans alike considered barbarians, who were the ancestors of the English – at least linguistically.

The Church in Rome sent out missionaries, among them St Paulinus. It was he who banished the fiends from Cross Fell, and later became Bishop of York. But for a while it was the monks of Ireland who were the chief proselytisers of northern Britain. In 633 Paulinus abandoned the North for Kent. The following year, King Oswald of Northumbria invited St Aidan, an Irish monk from the island of Iona, the cradle of Christianity in Scotland, to spread the Word among his people, and to build a priory. Aidan chose another island for his new priory: Lindisfarne. Perhaps he believed that islands would provide security from potentially hostile pagans on the mainland. Bede wrote that 'this island of Lindisfarne, with its little church and school, and its collection of rude huts, soon equalled, in the love and veneration of the northern English, the mother-foundation of Iona itself'.

Lindisfarne became the centre of evangelism in the north of what was emerging as England – although England was then no more than a collection of competing Anglo-Saxon kingdoms. The most celebrated figure among the Celtic monks at Lindisfarne was St Cuthbert, patron saint of Northumbria, who became prior around 665, and who ended his days as a solitary hermit on the nearby island of Inner Farne. Many miracles were attributed to Cuthbert, both during his life and after his death, and Lindisfarne became known throughout Europe as a centre of pilgrimage, learning and craftsmanship. Among the treasures created there was the Lindisfarne Gospels, one of the finest illuminated manuscripts in the world, dedicated to God and St Cuthbert. The Gospels are thought to be the work of a single monk called Eadfrith, who became Bishop of Lindisfarne in 698. The illustrations use a range of pigments from animal, vegetable and mineral sources, some imported, many local. Purples, blues and crimsons were made by adding an acid or alkali to various lichens or plant extracts such as woad and turnsole; reds and oranges were toasted lead; yellow was a naturally occurring arsenic-sulphide mineral called orpiment; green was verdigris; black was derived from charcoal; and white was chalk, or crushed eggshells or seashells, both widely available on the island. The pigments were bound together with gair (egg white), and the ink used for the text was made from oak galls and iron salts. The 516 vellum pages used the skins of some 150 calves, and were subsequently bound in leather decorated with silver and gold and jewels. The result is one of the greatest masterpieces of early medieval art. 'If you take the trouble to look very closely,' wrote the later chronicler Gerald of Wales, 'you will notice such intricacies, so delicate and subtle, that you will not hesitate to declare that all these things must have been the work, not of men, but of angels.'

The Lindisfarne Gospels are currently in the care of the British Library. Such is their importance to regional identity, there is a campaign to restore the Gospels to a permanent home in the Northeast.

~

It was still lashing down as I drove across the causeway to Lindisfarne, a couple of hours before the road would be inundated by the tide. A handful of pilgrims were completing St Cuthbert's Way, hobbling over the sands, the rain at their backs. Some of them carried staffs. I felt a fraud again – but a dry and comfortable one, sheltered from the elements in my car. If you walk across the sands between the mainland and the Snook, the landward end of Lindisfarne, you must follow a line of tall posts, and make sure you know the tide times. Drivers too can get into trouble, and can take shelter if necessary in a raised refuge box by the bridge over the channel of the little River Lindis as it meanders through the sands. It is this river that may have given the island its name. Lindis is from Old Celtic *linn*, 'pool', alluding to the waters that cut off the island at high tide; the second element in the name presumably alludes to the nearby Farne Islands. The island's other name, Holy Island, is that favoured by the Ordnance Survey, with 'Lindisfarne' written beneath it in Gothic script, indicating the name of an ancient monument.

Lindisfarne is so flat that all you can see of it from miles away – whether travelling on the East Coast Line or driving on the A1 – is the castle, sitting on its lump of dolerite. Much of the rest of the island comprises sand dunes below the first contour line and a few bare fields, ringed by tidal flats and stretches of rocky shore. Some of the names of the coastal features reflect the natural richness of the place: Primrose Bank, Snipe Point, Cockle Stone, Sheldrake Pool. There are still a handful of small fishing boats operating out of the harbour, mostly after crab and lobster. On a previous visit I had talked to a couple who farmed on the island, a husband and wife. They were selling their produce to the tourists at one of a number of roadside stalls. They grew barley, and had formerly had sheep, but they now let out the grazing. 'We grow all the fruit and veg,' the woman told me, 'to make the jams and chutneys we sell. And we grow our own cabbage, broccoli and beans. It's back-breaking work.' She began to get heated. They had, it seems, a rival. 'It really annoys us that the chap next door tells people he grows his strawberries here. What he actually does is buy them from the supermarket and bring them here in his van.'

'Oh dear,' I said.

'Never mind,' she said.

'He's clever,' her husband chipped in.

'He's greedy,' she said.

Fishing and agriculture are now only small parts of the island's economy. The big business is tourism. One of the most obvious visitor attractions is the fairy-tale castle that dominates the island's skyline. The castle is, however, something of a fraud. Even the original postdates the Middle Ages, having been built by Henry VIII as a modest fort – little more than an artillery platform – against incursions by the Scots. With the unification of the crowns of England and Scotland in 1603, the need for such defences largely disappeared, and the castle fell into disuse until

1901, when the ruins were purchased by Edward Hudson, the owner of *Country Life* magazine. Hudson commissioned the architect Edward Lutyens to renovate the place in the Arts and Crafts style as a holiday home. Lutyens, believing that every castle should have one, added a portcullis. The castle has been looked after by the National Trust since 1944, and has featured in a number of films, such as Polanski's *Macbeth* (1971), in which it stands in for Glamis Castle. It plays a larger role in the same director's psychological thriller *Cul-de-sac* (1966), in which an American gangster pushes his broken-down car across the causeway as the tide rises around him, then finds himself trapped on the island. He proceeds to take the owner of the castle and his young wife hostage, while awaiting instructions from his mysterious boss Katelbach. No word ever comes.

When I visited, the castle was half-covered in scaffolding, and the National Trust cash tills weren't working, so I got in for free. This was just as well, as there wasn't a lot to see apart from other tourists in their raingear, and some murky views over the sea.

I must have been in a mean mood. The rain didn't help. I had thought of visiting the twelfth-century priory, but English Heritage charge an entrance fee, and you can see most of the ruins without going through the gate. It was in these ruins –

> A solemn, huge and dark-red pile,
> Placed at the margin of the isle

– that Walter Scott imagined Lord Marmion's mistress, the nun Constance de Beverley, being walled up alive for breaking her vows. A century or more after Scott published *Marmion*, archaeologists found, built into a wall, a stone coffin containing a skeleton. The ruins you can see today are much more recent than the original foundation, destroyed in the year 793 in the first large-scale Viking raid on England.

Norse traders had been visiting Northumbrian shores for years, their intentions peaceful. But pressures back home – where population growth was outstripping agricultural production – led to a change of approach. Stories of the great treasures of Lindisfarne, an island easily accessible across the North Sea, defended only by unarmed monks, must have filtered back to the hungry pagan northlands. The *Anglo-Saxon Chronicle* later suggested there had been omens of the horror to come: 'In this year terrible portents appeared over the land of Northumbria, frightening the people. There were whirlwinds and flashes of lightning, and fiery dragons were seen flying through the air. A terrible famine followed.' On 8 June 793, a fleet of Viking long-ships drew up on the shores of Lindisfarne. Simeon of Durham, writing three centuries later, describes what happened next:

> They came to the church of Lindisfarne, laid everything waste with grievous plundering, trampled the holy places with polluted steps, dug up the altars and seized all the treasures of the holy church. They killed some of the brothers, took some away with them in fetters, many they drove out, naked and loaded with insults, some they drowned in the sea.

The event, the 9/11 of its day, was an existential shock to Christian England. 'Never before,' wrote Alcuin of York, 'has such an atrocity been seen. The heathens poured out the blood of saints around the altar, and trampled on their bodies like dung in the streets.' It looked like the light of Christendom was about to be snuffed out. Alcuin himself took the event as 'the sign of some great guilt' on the part of his fellow countrymen, a sign of divine displeasure. Two years later the Vikings attacked Iona itself.

Raids on Lindisfarne continued, until around 875 the monks finally abandoned the island, taking with them the Lindisfarne Gospels and the (apparently still uncorrupted) body of Cuthbert. It wasn't until the Norman era that a priory was re-established on the island. That in turn was dissolved, along with all the other monasteries in England, by Henry VIII. The island entered a period of decline, with the islanders reduced to relying on smuggling and shipwrecks, as reported by the seventeenth-century Catholic missionary Gilbert Blakhal:

> The common people there do pray for ships which they see in danger. They all sit down upon their knees and say very devoutly, Lord, send her to us, send her to us . . . They pray, not God save you, or send you to the port, but to send you to them by shipwrack, that they may get the spoil of her.

By the nineteenth century, fishing had become a flourishing industry, but not without cost. On just one day in 1865 five members of a single family drowned when their boat went down. Like other fishing communities, the islanders provided crews for the local lifeboat, which was responsible for saving 345 lives until it was taken out of service some fifty years ago. There is still a coastguard team based on the island, keeping an eye out for boats and drivers and pilgrims in difficulty.

Today pilgrimage to Lindisfarne has once again become an important part of the local economy. Notices in windows advertise all kinds of religious retreats, from 'Passing the Harp' and 'Finding Home', to 'Look at the Birds' and 'Celtic Advent'. Most of the retreats cost more than £200. Outside the medieval St Mary's Church, tit boxes and sparrow hotels were for sale in aid of church funds. The church itself was empty apart from an old lady rearranging the candles. She was, I supposed, waiting for the start of 5.30 prayers. As I left the churchyard an elderly priest limped past, dressed against the weather in Tilley hat, grey flannels and green waterproof. There would only be the two of them praying as the dark September afternoon turned towards dusk. Outside, martins massed and danced in the air above the graves.

In the early 1950s, when both the priest and his parishioner would have been in their prime, the author J. H. Ingram visited Lindisfarne for his survey of English islands.[*] He said he'd only

---

[*]   J. H. Ingram, *The Islands of England* (1952).

gone on a half-day's visit, but ended up staying three weeks. Ingram noted that many old customs still persisted on the island. There were a number of linguistic taboos, once common on board fishing vessels, but here also strictly observed on dry land. A pig could never be called a pig; it was 'the article' or 'yon thing'; if the animal was called by its name, listeners would reach out for a poker or other iron object. At weddings the bride would be assisted by two old fishermen to jump over the Petting Stone, the base of an ancient cross in the churchyard; if she stumbled, the marriage would turn out badly – but ill luck might be avoided if a plate was thrown over the bride's head and smashed to pieces on the ground. Perhaps because the tides play such an important role in the island's life, old people could be seen venturing out of their houses to curtsy to the new moon. Twenty years after Ingram was writing, another author[*] reported that certain customs and superstitions continued on the island. Some fishermen would not sit by their wives in church, and would refuse to set sail if they met a woman after midnight between home and harbour. They would also avoid meeting the parson in the street before noon.

Of the islanders in general, Ingram concluded that 'they are a law unto themselves, quick-tempered, ready with tongue and fist, but united against all outside interference'. Ingram doesn't say what brought him to this conclusion, but outsiders are eager, it seems, to identify the oddities of those who live on islands. I was no exception.

The next morning I called in at the visitor centre, run by the Holy Island Development Trust. There was a charge to see the centre's extensive and informative exhibition, but I was assured by the manager that all the money stayed on the island, to help fund affordable housing for locals. Unlike, she said, the money you paid to either the National Trust or English Heritage.

Before taking this job, she'd taught sewing to adults, and now used her skills to help mend the kneelers in the church. She was

---

[*] Donald McCormick, *Islands of England and Wales* (1974)

originally from Yorkshire, and now lived in Berwick, a few miles to the north. She told me she wouldn't want to live on the island. The islanders, she suggested, could sometimes be a little eccentric. She told me about the woman who hoovered the pavement in her nightie, and cleaned the insides of her windows wearing absolutely no clothes at all. Living here would be like living in a goldfish bowl, she said, 'with tourists always sticking telephoto lenses in your face'. Visitors could be very demanding. There was one, for example, who'd knocked on a neighbour's door and asked to use the lavatory. 'I don't use public toilets,' the visitor had explained.

There were two main categories of visitor, I was told. Firstly, there were the 'donkey tourists'. They'd demand 'Where's donkeys, where's beach, where's slot machines?' Then there were the pilgrims. 'A lot of them carry crosses made from a couple of fence posts nailed together.' One cross-carrying young man she'd encountered was dressed in nothing but a polyester gown. 'It was see-through in the wet,' she told me. 'I gave him some advice.'

'What was that?' I asked.

'Put on some pants,' she said, chuckling.

I paid my money, happy that it was in a good cause. After touring the exhibition, I still had time to spend before the tide receded sufficiently for me to drive off the island so, in the wet and the wind, I clambered up the dolerite outcrop above the harbour called the Heugh, on top of which stands the island's war memorial, also designed by Lutyens. Eight names are listed from the First World War (when the population of the island was around 600; it is now 180), and three from the Second. Some of the dead were fishermen whose boats were commissioned into the Royal Naval Reserve for such duties as minesweeping. One of the dead was a stoker on the battlecruiser HMS *Hood*; he was killed on 24 May 1941 on the far side of the North Sea along with 1,414 other crew members when the *Hood* was sunk by the *Bismarck*. Also on the Heugh is the former coastguard lookout tower. There are still islanders who remember volunteering here as lads to keep an eye on passing ships, at night and in stormy weather, and to raise the alarm if needed.

Even down in the harbour there was no escape from the weather. There were no fishermen, no other sign of activity, only a small team from Trinity House. They were here to ensure that the lights of Lindisfarne – one on the Heugh, the other a stone obelisk offshore on Guile Point – keep guiding mariners safely into the island's only sheltered haven.

The tide was now sufficiently low to allow me to re-cross the causeway. I had less than an hour to drive to Seahouses, a village a few miles down the coast, where I was booked aboard a boat for the Farne Islands. As I headed south I could see that the wind was beginning to stir the sea. The rain pitched down. When I reached Seahouses and walked head down to the harbour, a sign for Billy Shiel's Boat Trips told me that all sailings were cancelled. The seamen along these coasts have learnt, from bitter experience, that it pays to defer to the weather. The Farnes would have to wait for another day.

A Land Lost and Found

The Scillies

A land of old upheaven from the abyss
By fire, to sink into the abyss again;
Where fragments of forgotten peoples dwelt . . .

–Alfred, Lord Tennyson, *Idylls of the King* (1859–85)

Of all the cobwebs that nestle in the corners of the English imagination, those woven out of Arthurian whimsy have proved amongst the hardest to sweep away. I say 'English', although the Arthurian legends have their roots among the pre-English Celts – the Ancient Britons. They were later seized on and perpetuated by medieval romancers in France and Germany. But above all others it's the English – from Malory, Tennyson and the Pre-Raphaelites to T. H. White and contemporary pedlars of sword-and-sorcery fantasies – who have taken Arthur as their own.

Among the Arthurian cobwebs is the story of the lost land of Lyonesse, at various times identified with the district of Léonais in Brittany and the former kingdom of Lothian (called Leonais in Old French) in east-central Scotland. As the Arthurian legends evolved, Lyonesse became associated with Cornwall, specifically a land, now sunk beneath the waves, between Land's End and the Isles of Scilly. This sunken land has equivalents in neighbouring Celtic regions: the drowned city of Ys off Brittany, and, off the west coast of Wales, Cantref Gwaelod (the Lowland Hundred), a fabled kingdom now covered by the waters of Cardigan Bay. Like Atlantis, these places represent a Golden Age that is gone forever from the physical world, but which nevertheless live on as an embodiment of dreams and ideals – just as, in ancient Irish legend, Tír na nÓg, the 'land of youth', is the land far out in the western ocean where gods, fairies and the most blessed of mortals stay young, beautiful and healthy forever. With the coming of Christianity to Ireland, Tír na nÓg metamorphosed into the *Terra repromissionis sanctorum*, the promised land of the saints. It was in search of this isle of the blessed that in the sixth century St Brendan sailed westward; interpretations of *The Voyage of St Brendan*, an account written down several centuries later, suggest he may have reached the Azores, or the Canaries, or the Faroes. Some enthusiasts insist he reached North America.

Within the Arthurian legends Lyonesse existed as a solid, geographical location. For Malory, it was the birthplace of Tristan (whom he dubbed 'Sir Tristram of Lyones'), whose doomed love for his uncle's wife Iseult or Isolde led to the death of them both. For Tennyson, in his *Idylls of the King*, Lyonesse was the site of Arthur's final, fatal battle, against the treacherous Modred:

> Then rose the King and moved his host by night,
> And ever push'd Sir Modred, league by league,
> Back to the sunset bound of Lyonnesse –
> A land of old upheaven from the abyss
> By fire, to sink into the abyss again;
> Where fragments of forgotten peoples dwelt,
> And the long mountains ended in a coast
> Of ever-shifting sand, and far away
> The phantom circle of a moaning sea.

There are fragments of geological truth in Tennyson's account. Penwith (the western tip of Cornwall) and the Isles of Scilly are granite outcrops, surface exposures of the so-called Cornubian

Batholith, formed 280 million years ago by the crystallisation of an upsurging of magma – molten rock – beneath the surface of the earth. So the Isles of Scilly do indeed represent 'A land of old upheaven from the abyss / By fire'.

Many, even most, of the Scillies' fifty-five islands and islets may even in historical times have been joined together. The old Cornish name for the Scillies was *Ennor* (from *en moer*, 'the land' or 'the great island'), suggesting a single land mass. The Romans also referred to the Scillies in the singular as *Silina*. (The Roman name may be related to the same Celtic goddess who gave her name to *Aquae Sulis*, the Roman name for Bath.) There is evidence – such as old field boundaries now underwater – that the low-lying land linking the higher ground of today's islands was inundated by the sea some 1,500 years ago, probably as a result of post-glacial rebound. Post-glacial rebound is a slow process that has been altering the shape of Britain since the end of the last ice age. The disappearance of the ice cap that had lain heavy over much of the land resulted in the north of Britain, disburdened of the mass of ice, beginning to rise. The consequential tilting of the landmass meant that the far south of Britain, which had not been under the ice, began to sink. This process continues to this day, and southern England has lost much land to the waves. So the legend of a land that has sunk 'into the abyss again', as Tennyson puts it, may well echo a historical truth.

~

All the way through my children's childhoods we spent the summer holidays camping in Cornwall, at a place called South Treveneague, not far from Penzance, between Goldsithney and St Erth. Even though the site was away from the coast and sheltered by pine trees, we were often battered by high winds and rain, reminding us how exposed Cornwall is to the weather systems storming in from the Atlantic. On wet, windy days, the kids would build dens of fallen branches in among the bracken beneath the pines. Sometimes, if no grown-ups were about, they'd peer over the fences that ringed old mineshafts, while buzzards mewed anxiously in the treetops. When the weather was fairer, we'd spend days body-boarding on the surf beaches of West Penwith, or walking the coastal path, or, when the children were older, rock climbing on the sculpted granite sea cliffs of Bosigran, Chair Ladder and Pedn-mên-du, 'the black headland', near Sennen Cove.

From the top of Pedn-mên-du, by the old coastguard lookout station, we'd gaze southwest across the sea to the lighthouse perched on the sharp scattered rocks of Longships. Then we'd peer further into the distance in the hope of making out the Isles of Scilly. Sometimes we'd see one of the large blue and red Sikorsky helicopters that used to run a regular passenger service between Penzance and the Scillies. But of the islands themselves we saw no sign. I assumed they were shrouded in mist or haze, or hidden below the horizon – or just some figment of England's collective imagination, now sunk beneath the surface of the sea.

Then, on an early evening in August, on the last family holiday we had together in Cornwall, my son and daughter joined me at the top of a climb up the steep granite flakes and whorls of Pedn-mên-du. The air was still, the light clear and there, far beyond Longships, among the pink and grey streaks of the western sky and the western ocean, appeared a pattern of rocky islets. They could have been set on the surface of the sea, or they could have been floating in the air. They were undoubtedly real, and yet completely unattainable. It was my first sight of the Isles of Scilly.

Pedn-mên-du is just north, and barely east, of Pedn-an-Wlas ('headland of land'), better known as Land's End, the most westerly point of the English mainland. The name 'Land's End' is first recorded in 1337, and echoes similar names in Spain and France: both Cape Finisterre in Galicia, and Finistère at the western tip of Brittany derive from the Latin *finis terrae*, 'end of the land'. The namers of these headlands, when they reached them, all concluded that they had come to a full stop. The earth was at an end; there was no place further to go; only the ocean remained, stretching out indefinitely westward. But those who looked out across the Atlantic, towards where the sun sets, conceived of places beyond the horizon where the sun still shone even when their own land was shrouded in night. In Irish legend, beyond the Aran Islands off the west coast, lay Hy-Brasil, one of the refuges of the Tuatha Dé Danann, the old gods of Ireland, once they had been deposed by new waves of invaders, and denied their divinity by the early Christian missionaries. Hy-Brasil was for long regarded as a reality, and was marked on old charts. According to one theory, when the Portuguese landed in South America in 1500 and named the easternmost part of it 'Brazil', they were thinking of the old Irish legend.

In Breton lore, beyond Finistère, the sunken city of Ys may, like Hy-Brasil, have represented the lost culture of the pre-Christian Celts, drowned out by the new dispensation. The Welsh legend of Cantref Gwaelod also recalls such a lost past, although the land that is lost may have been submerged long before the Celts – let alone the Christians – even arrived on these islands. At Ynyslas on Cardigan Bay, the stumps of an ancient forest of oak, pine, birch, willow and hazel can be seen at low tide. These remains are some 5,000 years old, and are evidence of a habitat that would have provided hunter-gatherers with rich pickings.

The fable of Lyonesse also echoes the real drowned land of Ennor that once linked the now-scattered Scillies. There is some evidence of a Neolithic presence on the island(s), but farmers and fishermen do not appear to have settled there permanently until the Bronze Age. Presumably human pressure on the resources of West Penwith must have been building when some visionary looked out from the end of the land one clear day and saw in the distance another, unknown place that might provide harvests from both earth and sea. It must have taken some courage, or desperation, to set off in dugout canoes or leather-clad coracles across thirty or so miles of exposed ocean.

It is said that there are more Bronze Age barrows – tombs – on the Scillies than in the whole of Cornwall. A good number of these tombs are passage graves, in which a stone-lined tunnel leads to a burial chamber under an earthen mound. Here the cremated remains of the dead have been interred – or, more unusually, an unburnt skeleton has been placed with its knees tucked up under its chin. The plethora of burial sites led some people to speculate that the Scillies were considered by mainlanders as the Abode of the Dead, a sacred burial ground, the Cornish equivalent of the Welsh island of Bardsey, 'the Island of 20,000 Saints'.

Some in the past identified the Scillies with the Cassiterides, the 'Tin Islands' mentioned by the ancient Greeks and Romans (Greek *kassíteros*, 'tin') that supposedly lay somewhere off northwestern Europe. The Scillies have little in the way of tin, unlike mainland Cornwall, but may possibly have been a staging post for the import into southern Europe of the metal, one of the ingredients (alongside copper) of bronze, the dominant material between the ages of stone and iron. The tin trade was at first dominated by the Phoenicians, and then by the seafarers of Gades (modern Cadiz), who kept their sources a secret.

The first firm historical reference to the Scillies, '*insula Sylina, quae ultra Britannia est*', dates from AD 384, in late Roman times, when two bishops, Instantius and Tiberianus, were exiled there from Spain for upholding the heresy known as Priscillianism. Priscillianism, an ascetic belief system founded by an Iberian nobleman called Priscillian, was based on Gnostic-Manichaean dualism, which divided the cosmos into a Kingdom of Light and a Kingdom of Darkness, both of which were embodied in human beings. Nature and the material world belonged to the Kingdom of Darkness, hence believers followed a path of strict self-denial. Priscillian himself was executed for sorcery. Whether exile on the Scillies reinforced the bishops' belief in the darkness of nature and their own inner light is not known.

~

Some ten years passed between that first sight of the Scillies from a cliff top near Sennen and my first visit to the islands. I was travelling with my wife Sally and two old friends, Alice and Tom, who'd been to the Scillies before and were keen to return.

We almost didn't get there.

'Sorry,' said the voice on the other end of the line. I'd been told to phone the Operations Department at Land's End to check on our flight from Exeter, from where you can now fly to St Mary's, the largest of the Scillies. 'The problem's fog,' I was told. It was all over Cornwall, all over Devon. Fog at Newquay, fog at Exeter, fog at Land's End. 'We didn't fly this morning,' the man said. 'And your flight this afternoon might not happen.'

I asked what the alternatives were. 'We could bus you down to Penzance tomorrow morning, leaving Exeter at six. Then we'd put you on board the *Scillonian*.' The *Scillonian* is the passenger ferry to St Mary's. 'The *Scillonian*?' an old Scillies hand had repeated when I mentioned the ship. 'Ugh. Rolls like a pig. You'd better pack some seasick pills.'

'But who knows, we might fly this afternoon,' the man at the Operations Department continued. 'We won't know until the time comes. It'll be the pilot's decision.' I said we'd catch our train, and see what happened when we got to Exeter Airport.

We left Paddington under cool high cloud. Further west, the sun broke through. As the train cut south through Devon towards Exeter, we could see the remains of the fog. It was lifting around the hills, and was mostly now above them, but the threat of it still lurked in the still air.

Uncertainty still hung over Exeter Airport. 'We'll tag your luggage and hope for the best,' the helpful woman at the check-in desk explained. 'We've just heard the plane's taken off from Newquay. It should be here before too long. But the pilot may not land if it closes in again. And if he *does* land, he may not take off again. And even if he does take off again, he might get to St Mary's and decide he can't land there. Then he'll fly you back here. That's what happened on Tuesday. There hasn't been a flight to the Scillies all week.'

There was nothing to do but sit and wait.

After some time we heard a buzz in the still air, then saw a small, twin-engined plane coming in to land. It was our Twin Otter, only a few minutes late. It taxied up beside the small terminal building. But still we sat in the waiting room. We were shown the safety video, so we'd be ready to go if we were going to go. But, we were told, that wasn't guaranteed.

The tension mounted. Then someone thought they saw our bags being loaded onto the plane. Our check-in woman opened the door of the waiting room and ushered us out onto the tarmac. 'Best get on board quickly,' she said. 'Before he changes his mind.'

'Mind your heads,' said the pilot as he ushered us up the few steep steps into the fuselage. The seats were narrow, with low backs. There were singles down one side, doubles down the other, with a large crate for pet dogs somewhere in the middle. I sat at the back, beside one of the emergency exits. The pilot showed me the flap to lift to find the handle should we have to use it. Apparently this was my responsibility. I was puzzled by his accent. 'Nepali,' he said with a grin. I was reassured. If you can fly a plane between the giants of the Himalaya, then landing on a small rock thirty miles out in the Atlantic shouldn't be too much of a problem. As the pilot resumed his seat in the cockpit, we could see him and his co-pilot at their controls. There were no sealed doors to keep out hijackers on this plane.

The engines began to hum, the propellers to spin. The pilot released the brakes and we roared forward. I could see the runway dipping down ahead. Then, after what seemed like only a hundred yards, the nose was up and we were airborne.

The pilot told us we'd be flying at 8,000 feet – much closer to the ground than the 35,000 feet of the average passenger jet. At first we rose through cloud, but soon we were looking down at the remains of the fog. The tatters parted to reveal a dense and complex patchwork of fields, each one a different shape, a different shade of green or brown, dictated by the folds and curves of the land rather than by a ruler on a map. My view was framed by the Twin Otter's wing strut above and the non-retractable under-carriage wheel below.

A friend used to tell me that when driving down to Cornwall from London the light would change as he crossed Bodmin Moor. We noticed the same thing at 8,000 feet. The air became more transparent, brighter, and at the same time milkier. It is a strange paradox. The milkiness may be due to the fine spray hoisted into the air by the incessant surf battering Cornwall's wild coasts.

Looking down, I began to recognise some of the beaches where we'd spent summers with our children: Godreavy, Gwithian, White Sands, Sennen Cove. Then we turned southwest past Land's End, and were over the open sea, the place where Lyonesse was lost. After only a few minutes over the Atlantic, small shreds of land edged by curves of sand began to appear. The silvery turquoise water around the islands was shallow, barred by dark stripes of kelp: the sea was patterned like a mackerel. The Twin Otter dipped its nose, once more entering a realm of mist. Then we hit the short, hill-top runway and the plane braked hard to a stop.

We were driven down the hill to the harbour through Hugh Town, the main settlement on St Mary's, itself the most popu-lated of the Scillies (over 1,700 permanent inhabitants in 2011, ten times more than Tresco, the next most populated). Hugh Town is not unlike other small, busy Cornish harbour towns, its buildings a mixture of the picturesque and the prosaic. Alice had told me that for the wilder, more particular Scilly flavour we should stay on Tresco. She had arranged accommodation there, at the New Inn.

At Hugh Town harbour we were put aboard *Hurricane*, a small jetboat. The three-mile trip to Tresco only took a matter of minutes. The sky was dull with the memory of several days of fog, the air still. To the west, rock after rock jutted out of the shallow sea. Any sailor round these parts needs to know their charts and tides.

Our destination was New Grimsby, the main settlement on Tresco. Clambering up stone steps onto the jetty, we were met by an open-backed van called, according to its number plate, 'Dave'. Dave took us and our luggage the few hundred yards to the New Inn.

Dave was painted dark green. So were all the signs we saw, helpfully directing visitors to places such as the Island Office, Old Grimsby, Tresco Abbey Gardens and so on. There was another identical van, called 'George'. The only other vehicles on the island that we encountered were small golf buggies, all painted an identical green. I later found out the whole of Tresco, including its main enterprises (the New Inn, two other restaurants, a large shop and post office, holiday cottages and timeshare apartments), are all owned and run by the Dorrien-Smith Estate, which in turn has a long lease on the island from the Duchy of Cornwall (from which Prince Charles draws an income of some £20 million per year). The Duchy of Cornwall, although owning only around 2 per cent of mainland Cornwall, owns almost all of the Scilly Isles. Attempts by leaseholders on the islands to buy the freehold of their properties have been resisted by the Duchy, which benefits from an exception in the Leasehold Reform Act of 1967, a statute that allows most leaseholders who meet certain qualifying criteria to buy the freehold of their properties. The Duchy itself states on its website that it 'does not accept applications to enfranchise in certain specific geographic locations . . . [including] the off-islands and the Garrison area of the Isles of Scilly'. One leaseholder on St Mary's told the *Guardian*, 'The situation is feudal, unfair, lacks transparency, removes the legal rights of individuals and ignores equality.'

The Dorrien-Smith family, current proprietors of Tresco, owe their position to Augustus Smith, a Hertfordshire landowner who was also heir to a banking fortune. Smith acquired the lease on the entire Isles of Scilly from the Duchy of Cornwall in 1834. He declared himself 'Lord Proprietor of the Scilly Islands', a title also borne by his heirs until 1920, when the lease on all the islands except Tresco reverted to the Duchy.

Augustus Smith was not altogether popular. He banned subletting on the smaller islands, forbade young couples from marrying until they had a home of their own, and expelled any person who could not obtain local employment. In 1855 the ten inhabitants of the tiny island of Samson, who hitherto had subsisted on a diet of limpets and potatoes, were evicted to make way for deer (an echo of the Clearances then being carried out on the islands of the Hebrides far to the north). Smith did make some improvements, however: he built a new quay at Hugh Town, and introduced compulsory elementary education some decades before it became mandatory on the mainland. Pupils were charged a penny a week if they attended, and tuppence if they did not.

Smith left no legitimate heir (although he is believed to have fathered several children on his female servants). The estate was inherited by his nephew, Thomas Algernon Smith-Dorrien-Smith, and is now in the hands of Robert Dorrien-Smith, said by the *Telegraph* to be 'a close friend of the Prince of Wales'.

There is a monument to Augustus Smith in St Mary's Old Church on the island of St Mary's. The text of the plaque reveals much by what it does not say:

> This monument is erected by the inhabitants to preserve among them the recollection of a name henceforth inseparably connected with these isles.

A Cornish Post Office official in Penzance called J. G. Uren was more forthright in his 1907 book *Scilly and the Scillonians*: 'In all things appertaining to the government of the islands, his rule was absolute and his word was law . . . To oppose the Governor spelt ostracism and deportation.'

It was not for nothing that Smith was nicknamed 'Emperor'. He chose not to live on St Mary's, where most of his sub-tenants lived, but on Tresco, which he (and his successors) have turned into what is effectively a private island. (Ironically, Smith was a champion of the right of every Briton to walk on common land, and paid a team of navvies to dismantle the fences erected on

Berkhamsted Common by his Hertfordshire neighbour, Lord Brownlow of Ashridge.)

~

Our first evening on Tresco (from Cornish *tre*, 'farm', and *scaw*, 'elder trees') I chatted to some of the staff at the New Inn. The man behind the bar had previously managed a ski resort in the French Alps. He hadn't yet spent a winter on the island, but looked forward to it with equanimity. One of his colleagues was from Bath. She'd travelled the world for seventeen years before ending up on Tresco. She'd survived her first winter. She told me there was only one native islander left, a member of a family called Christian that had once farmed here. He now hires out jet skis, she said. This seems at odds with the claim on the official website of the Tresco Estate, which asserts that 'A community of about 150 people permanently lives on the island, with a mixture of young and old alike. Some families have lived on the island for many generations.' Later I spoke to a woman in the Estate office. She'd lived on the island for six years, and loved it. She told me winter was a relief after the busy tourist season. 'During the storms it's just cosy,' she said. We talked about the relations between the inhabitants, the Estate and the Duchy. Did she consider herself and her colleagues to be feudal underlings, I teased. She laughed. 'Not underlings. Tenants,' she said.

After a short time on Tresco – which bills itself as 'the luxury island resort' – I had the impression that there were two main groups of people on the island: the well-heeled, older visitors, and the young staff who serve them in the hotel and restaurants, and help to run the holiday accommodation. For the most part, it appeared that these young people worked on Tresco for just a few months, or perhaps a year or two, before moving on. On Saturday night those who weren't working seemed to congregate in the bar of the New Inn. I wondered what tensions might arise amongst a group of young people thrown together on such a small island – Tresco only measures two miles long by, at most, a mile wide. In

2015 the body of one of the bar staff working on the island was found by a French yachtsman on rocks off the small uninhabited island of Tean, northeast of Tresco, ten days after he had disappeared. One witness said he had seen the young man at a party, involved in an altercation. However, the police concluded that there was no evidence of foul play. The man's family remained unconvinced. The *Telegraph* headlined its report on the case 'Death in Paradise'.

~

While we ate our breakfast at the New Inn we looked westward across the narrow stretch of shallow water separating Tresco from the smaller, mist-covered island of Bryher. At low spring tides it is possible to walk (or at least wade) between Tresco and Bryher, and also to the small uninhabited island of Samson, off Bryher's southern tip, which Augustus Smith had cleared of its people back in 1855. (Samson is named after a sixth-century Welsh saint, Samson of Dol, one of the founder saints of Brittany, who at one point visited the Scilly Isles.)

It was October and there was no low spring tide so, rather than wading, we opted to take the small boat ferrying a handful of tourists to Bryher. The boat, larger than *Hurricane*, was called *Voyager of St Martin's*, after the most northeasterly of the inhabited Scillies. It would continue from Bryher to St Mary's, and then pick us up at lunchtime, before returning to Tresco. These small boats only confirm they are sailing the evening before, the passages between the islands being dependent on tide and weather. We would only have two or three hours on Bryher, but that was more than enough time to walk round the island.

As we paused on our way up what passes for a road on Bryher, through the small settlement that calls itself the Town, we were overtaken by a single file of elderly birders. They trudged past in a determined manner, head-down and silent. Ornithology is not a hobby for the talkative. They carried all the expensive kit that an affluent retired enthusiast could afford. Many were laden with tripods, and one had a lens the size of a traffic cone, carefully wrapped in camouflage material. We later saw a line of them standing stock still, peering through their binoculars at something we could not see. In the autumn the Scillies attract many rare migrants passing through, and even some strays blown in from the Americas. A couple of days later, on the island of St Agnes, I did manage to get an answer from one of the chattier birders, when I asked if he'd spotted anything interesting. 'Red-backed shrike,' he said. There was, apparently, no need to elaborate. It turns out that the red-backed shrike, once a common migratory visitor to Britain, had virtually disappeared from these shores, so a sighting was (although you'd never guess it) the cause of some excitement.

Back on Bryher we continued through veils of mist across an isthmus, and so up Samson Hill, a low gorse-covered eminence at the south end of the island. The Ordnance Survey marks 'Cairns' on the summit, in fact prehistoric chambered tombs. To the untutored eye, they seem no more than great blocks of granite. Unfortunately, the ancient aura of the place has been marred by a square black slab cemented into the rocks, ostensibly to celebrate the millennium in the year 2000. That date would seem irrelevant and that small lapse of time the blink of an eye to any shades from the distant past who might still linger here, attending to the bones or ashes they'd left behind, buried beneath the massive rocks. Who knows what names they might have had for the hill before Samson of Dol came here? Their language is unknown, predating the arrival of the Celts in these islands by hundreds if not thousands of years. The names that do remain on Bryher (itself probably from an Old Celtic word meaning 'the hills') suggest stories from a more recent but still long-gone past. There is no one left to tell the tales of Droppy Nose Point, Stinking Porth, Moon Rock, Popplestone Neck, Puckie's Carn, Hell Bay or House of the Head. The origins of some names *are* known: the grimly named Hangman Island, in the sound facing Tresco, refers to the story that Admiral Blake, who captured the Scillies for the Parliamentarians in 1651 (three years after the Royalists had been defeated on the mainland), built a gibbet here.

From the summit of Samson Hill we gazed out westward over a maze of rocks looming out of the sea into the mist: Gerwick, Buzza, Illiswilgig, Middle Ledge, Stippit, Maiden Bower, Mincarlo, Biggal, and many more, all of them named. The intensity of the naming attests to the dangers of navigating these waters, and the life-and-death necessity of knowing which rock was which. Some of the rocks resemble great fat seals lumbering onto land, others the backs of sea monsters rolling in the waves. Close up on Bryher we could see that every outcrop of granite was shaggy with lichen.

The change of the tide dictated that our boat back to Tresco embarked from a different quay than that on which we had disembarked. This first quay was now stranded high up its beach. Back on Tresco I used the afternoon to explore the northern part of the island, which is as rugged and granite-cragged as Bryher. Opposite and a little north from Hangman Island I came to Cromwell's Castle, a round tower built on a rocky promontory above the sea in 1651 after the Parliamentarian takeover. It was built to guard New Grimsby Harbour, the sound that provides a sheltered anchorage between Tresco and Bryher, and was essentially a gun platform. It continued to be manned and modified through the following century. The unfortunate men who were stationed here, in what must have been a draughty and damp building, relieved their boredom by inscribing dates and initials into the stone and plaster of the interior: 'T.S.E.' was here, as was 'F.P.E.' and 'E.J.' The year 1761 slowly came, and as slowly passed.

Cromwell's Castle superseded King Charles's Castle on the hilltop above. King Charles's Castle was actually built during the reign of Edward VI (1547–53), long before Charles I ascended the throne. King Charles's Castle was attacked by Blake's Parliamentarians in 1651, but the defenders succeeded in blowing it up before the attackers took complete possession. There is now little left bar a few low walls and some massive granite lintels.

Even in October there was bell heather still in flower up on the heath, even the occasional sea pink, set off by the yellow of hawk-weed. A stonechat chatted, a rock pipit bobbed from rock to rock, while oystercatchers hurried across the millpond water of the sound between Tresco and Bryher as the fog rolled in and out. Further north, as the sound opened out into the exposed Atlantic, the ocean swell was breaking over the rocks of the Horse and Shipman Head over on Bryher. I could hear it from half a mile away through the still air. And so I turned round Kettle Point and Gun Hill at the northern tip of Tresco, and made my way, past Piper's Hole, down the east coast to Old Grimsby, the small settlement on that side of the island. The wild windswept heath of the northern half of the island was replaced by green fields, palm trees and exotic, subtropical succulents.

Walking in the lessening light across the neck of land between Old and New Grimsby I came to the island's only church, a modest exercise in Victorian neo-Gothic. The interior was lit through stained-glass windows, enough to make out the words on a wooden plaque commemorating Geoffrey Richard Dorrien-Smith of the Parachute Regiment, who had been killed at Arnhem on 21 September 1944 at the age of twenty-eight.

As I left the church an old man entered. He nodded to me shyly as he passed. After I'd shut the door behind me and began to explore the churchyard outside, there was a sudden outburst of piano music from within the church – densely notated Romantic stuff, fevered and frantic and slightly out of tune, but played with considerable virtuosity. I must have been the only live listener. Among the attentive dead who lay about me were William Alfred Nicholls and his brother John Hope Nicholls, both of whom had died in France in 1916. Of those who had perished closer to home I found Samuel Jenkins, who had drowned on 17 February 1879, at the age of fifty-eight years. Jenkins was at one time the commonest surname on Bryher. On Samuel Jenkins's tombstone there was an inscription from Psalm 68: 'I will bring my people again from the depths of the sea.' I assumed that Samuel Jenkins's body had been recovered from the depths of the sea, so that it could be buried in this peaceful churchyard. Around me the muffled, off-key music of the church piano swirled in the dying light.

~

The following morning the rain came, and with it the wind. The *Scillonian* did not sail that day. We opted for Tresco Abbey Gardens, the sheltered subtropical oasis in the south of the island.

No one was about as we walked up the drive to the grand pile of Tresco Abbey – not an abbey at all, but a nineteenth-century extravaganza, half Germanic castle, half Italian palazzo, designed and built by Augustus Smith himself, and expanded and elaborated by his successors. It was Augustus Smith who started the gardens, gouging terraces out of the hillside and planting shelterbelts to create a habitat where many Mediterranean and subtropical plants could thrive away from the wind, washed by the balmy airs of the Gulf Stream. Smith, recognising the botanical potential of his new, mild fiefdom, introduced plants from South Africa, California, Mexico, Australia and New Zealand. Smith's work was carried on by his nephew, Thomas Algernon Smith-Dorrien-Smith, who understood the commercial potential of the Scillonian climate, which allows narcissus to flower on the islands a month earlier than anywhere on the mainland. Algernon bought a steam launch to export this profitable crop to the rest of England. Algernon's son Arthur travelled widely to collect plants and seeds, an interest inherited by *his* son Thomas, a lieutenant-commander in the Royal Navy, fondly known as 'Commander Tom'. Commander Tom gave many lectures about his passion, and would lead his audience on imaginary walks around the gardens back on Tresco. 'It is a man of poor imagination', he would declare, 'who cannot see the gibbons swinging through the tall trees and the anacondas lazily hanging with good appetites. I myself have frequently seen a tiger's tail disappearing into the astelia undergrowth.'

At the entrance to the gardens there is an old slab of slate bearing an inscription requesting visitors, among other things, 'to abstain from scribbling nonsense and committing suchlike small nuisances'. The weather being what it was, there was no one about in the gardens to commit any kind of nuisance. Above us, the heads of tall palms nodded in the wind. Even in October there was much in flower – but there were no gibbons swinging, no anacondas hanging, not even the faintest swish of a tiger's tail.

Although most of the Smith-Dorriens' botanical introductions have been contained within the gardens, a few have broken loose, and are altering the local ecology. In many parts of the islands we visited, among the native wild flowers, there were swathes of red-hot pokers and pink nerines, matts of Hottentot fig and, everywhere, the sharp, sword-shaped leaves of agaves. In addition, on the sheltered eastern side of Tresco, there were thickets of rhododendron – the invasive, mauve-flowered *Rhododendron ponticum* that has throttled out native flora and fauna in many parts of western Britain, from Snowdonia to Torridon. Elsewhere, major efforts are being made to deal with such invaders. Along the Cornish sea cliffs much work has been done to remove carpets of Hottentot fig, with its spear-shaped succulent leaves, and its spectacular, shocking-pink flowers. On the island of Lundy, after many years, the wardens have eventually managed to extirpate *Rhododendron ponticum*. On the Scilly Isles, other invaders have also proved a problem, most notably the brown rat, which has wreaked havoc with breeding seabirds. In 2013 an eradication programme was instituted on St Agnes, where the populations of Manx shearwaters and storm petrels had declined by a quarter in twenty-five years. By 2016 the island was declared rat free, and the birds have begun to recover. However, one of the boatmen I talked to told me that the programme, if it is to be extended to the other Scillies, requires 100 per cent of an island's population to agree. He said this wasn't happening. There was always a handful of people who felt a sentimental attachment to the rats. On Tresco, we saw more than one fat, sleek rodent slinking off

into the hedgerows. But we also saw red squirrels, recently rein-
troduced to the island. They had been flown in by Royal Navy
helicopter.

~

We emerged from the dark, subtropical shelter of the gardens
into a blustery, bright autumn afternoon. Clouds and showers
scudded overhead as we walked to Tresco's southern shore. Here,
backed by dunes and separated by craggy granite headlands,
white sands slope down to the clear turquoise channel separating
Tresco from St Mary's. It could have been the Adriatic. As we
walked along by the line of washed-up bladderwrack marking
the top of the tide we set off flight after flight of ringed plovers.
The wingbeats of these small wading birds were almost too fast
to see as they described sharp anxious circles low above the waves.
In a moment they were back on shore, several more yards ahead,
their nerves settled, picking at creatures too tiny for us to see. As
they became accustomed to our presence, they would delay their
next quick flight until the last possible moment. Further along
the shore some gulls took to the air, their wingbeats slow and
ponderous in comparison.

As we turned the island's southeastern corner, by Diamond
Ledge and Tobaccoman's Point, the weather closed in. The sky
turned dark, rain squalls rushed in from the west, and above
the white horses flecking the waters offshore a rainbow spanned
the channel between Tresco and St Martin's. God's promise to
Noah had come too late for Samuel Jenkins, buried in Old
Grimsby churchyard, too late or quite forgotten for all the others
who had drowned in the waters around these islands.

The Scillies have a turbulent history of wrecks, long lists of
deaths at sea. On 22 October 1707 the islands witnessed one
of Britain's worst ever maritime disasters, when several ships of a
Royal Navy fleet sailing from Gibraltar to Portsmouth under the
command of Sir Cloudesley Shovell were smashed against the
Western Rocks, between St Agnes and the open Atlantic. Four

ships were lost, and at least 1,400 men were drowned. For days afterwards, bodies and wreckage were washed up on the inhabited islands. The corpse of Shovell himself came ashore on St Mary's. A legend grew up that he was still just about alive when he landed, but was finished off by a local woman for the sake of his emerald ring.

The lost ships of 1707 were just a few of the hundreds of ships that have come to grief around the Scillies. One of the greatest tragedies occurred on 7 May 1875, when the SS *Schiller*, a steamship carrying passengers from New York to Hamburg via Plymouth and Cherbourg, encountered thick fog as it approached the Scillies. The captain asked for volunteers to keep an eye out for the lighthouse on Bishop Rock, the westernmost of the islands' many perils. He offered a bottle of Krug champagne for the first to spot the light, or hear the fog bell. To no avail. At ten o'clock that evening the *Schiller* hit the Retarrier Ledges, an outlier of the Western Rocks. The ship had passed inside the unseen Bishop Light, into the danger zone. The captain ordered the engines be put into reverse, to back the ship off the reef. The manoeuvre was successful, but the rising swell simply threw the *Schiller* onto the rocks again, this time sideways. Panic broke out on board. Some of the male passengers wielded knives in an effort to secure a place in the lifeboats. The captain fired his pistol in the air to restore order, and when he ran out of bullets he drew his sword to force some of the men out of the lifeboats. All the time the grounded ship was being struck by giant breakers. Two of the lifeboats were crushed when one of the ship's funnels fell on them; two more capsized; and another two were smashed to pieces against the hull. The women and children remaining on board had taken shelter in the deck house, but to the horror of onlookers, a huge wave smashed off its roof, and another swept all the occupants into the sea. Only twenty-seven people managed to escape in the two lifeboats left intact; a few other survivors were eventually picked up; the rest of the passengers and crew, 335 people in all, perished.

Wrecks have always had an upside for those on shore. Goods washed up have in the past helped to sustain impoverished coastal communities. There is little or no evidence for stories of 'wreckers', locals along the shores of Devon and Cornwall who would supposedly show false lights to lure ships onto rocks, and then steal their cargoes. But if there was a wreck, there might also be a profit. Towards the end of the eighteenth century, the Reverend John Troutbeck, chaplain of the Isles of Scilly, reputedly offered the following prayer: 'We pray thee, Lord, not that wrecks should happen, but, that if any wrecks should happen, Thou wilt guide them into the Scilly Isles, for the benefit of the inhabitants.' Today, such activities would be called salvage. (The Reverend Troutbeck was dismissed from his post in 1795 for handling smuggled goods.)

Although Augustus Smith, first Lord Proprietor of the Scillies, was wealthy enough not to have to rely on the perils faced by passing ships, he nevertheless kept an eye out for wrecks around his domain. His particular interest was ship's figureheads, which he installed in his gardens in a specially built museum he called 'Valhalla' (the home of the gods and the heroic dead in Norse mythology). The collection was continued by his successors, who added the figurehead of the *Schiller*. In a similar vein, on Lundy the inn displays lifebelts from some of the many ships that have been wrecked around the island's shores. Are such displays unhealthily morbid, like trophies of war? Or are they rather an acknowledgement of the dangers faced by those who, in the words of Psalm 107, 'go down to the sea in ships, that do business in great waters'?

Over the centuries, the people of the Scillies – especially the many skilled pilots – have been responsible for saving hundreds of lives of those whose ships have come to grief around their shores. Their lifeboats were six-oared pilot gigs, capable of running unharmed before huge seas. Today, competing pilot gigs regularly race each other every summer, the world championships being held in the Scillies. The Ordnance Survey even marks one of the courses, between St Mary's Pool at Hugh Town and Nut Rock off Samson.

Ownership of the Valhalla collection on Tresco was transferred to the National Maritime Museum in 1979.

~

That evening a notice was posted at the New Inn that told us the next day there would be a boat to St Agnes. But we would only have two or three hours on the island before it returned again to Tresco.

The waters were calm as we sailed south down the shallow sound between Tresco and Bryher. But once we emerged into the Road, the deeper channel northwest of St Mary's, we were exposed to the full swell of the Atlantic. Rows of cormorants stood at attention on each of the many rocks we passed; drifts of cormorants spread across the surface of the sea. When an unseen shoal of fish passed beneath, the birds would roll as one and slip under the surface. And so we passed Great Rag Ledge and Nut Rock off Samson, then Doctor's Keys and Woolpack Point on the west side of St Mary's, until we came to Cow and Calf, and Kallimay Point guarding the entrance to the bay on the north side of St Agnes called Porth Conger, where we finally docked at the quay.

Although the church on the island is dedicated to St Agnes, the name of the island has nothing to do with her. It was formerly plain *Agnes*, from Old Norse *hagi*, 'pasture', and *nes*, 'peninsula', referring to a promontory on the island; the *Saint* element was later added to chime in with the neighbouring islands of St Mary's and St Martin's.

St Agnes has a conjoined twin called Gugh, pronounced 'goo', to which it is semi-attached by a spit of sand called the Bar. The Bar is close to the quay where we landed, and, although the tide was rising, there was just time to make a quick visit to Gugh. On the St Agnes shore there is a sign that warns of 'strong currents when sandbar is covered'. It wasn't, so I hurried across. Behind me I was aware that the sea was cutting a channel into the sand of the Bar, so as soon as I reached the Gugh shore I turned round and headed back to rejoin my companions.

So I missed the Neolithic entrance grave on Gugh called Obadiah's Barrow, up on the slopes of Kittern Hill. Obadiah's Barrow was excavated by the antiquary George Bonsor in 1901. Within the burial chamber, originally covered by six large capstones, he found a disarticulated human skeleton, probably late Neolithic, and a dozen cremation urns from the Bronze Age. It seems the site, its entrance facing west towards the setting sun, was used for burials over a period of many hundreds of years. The current name of the site indicates a colonisation of an ancient past by the Victorian antiquary, Obadiah being the name of several figures in the Old Testament, including a prophet. None of these characters had anything to do with the Isles of Scilly.

There was, however, a local man called Obadiah. Obadiah Hicks was a St Agnes pilot who early in the morning of 8 May 1875 had taken his gig out to investigate reports of cannon shots beyond the Western Rocks. A cannon shot indicated a ship in distress. As Hicks and his crew approached the Retarrier Ledges they could make out the masts of the *Schiller*, with men clinging to the rigging and bits of wreckage. Those who still had the strength were screaming. The gig was shipping water in the heavy seas, and when its rudder was damaged by floating wreckage, Hicks was forced to turn round and seek safety on St Mary's. He managed to pick up five men from the water on his way. They only just made it, the crew having to bail furiously to stop the gig from foundering. The mail steamer *Lady of the Isles* and a number of small boats took over the work of rescue. But it was largely a work of recovery. Most of the dead were buried in three mass graves on St Mary's. The islanders had to blast away the granite with dynamite to make room for all the corpses.

As most of the passengers had been German, the Kaiser sent tokens of gratitude to the islanders, in thanks for their efforts to save his countrymen. There was a rumour that throughout the two world wars, both the German navy and the Luftwaffe were ordered to avoid attacking the Scilly Isles, as an extension of this gratitude. This seems unlikely. Although the sheltered waters round the Scillies were too shallow to provide an anchorage for modern British battleships, Tresco was used in the First World War as a seaplane base for anti-U-boat operations, and in the Second World War as the base of a flotilla of small, fast naval vessels disguised as Breton fishing boats, sent across the channel to carry agents in and out of France, and to gather intelligence about German coastal defences in Normandy in the run-up to D-Day.

St Agnes has a different feel to either Bryher or Tresco. It is more a working island than a holiday destination, with a population of around eighty. A blackboard on the community noticeboard announced the recent birth of twins to islanders Kate and Rob. In the disused Methodist chapel, now a community centre, I fell into conversation with a local weaver called Trish, who had a studio there. She told me there was no hotel on the island, and only two second homes. There were farm tenancies, and some of the islanders had been making a living from fishing for three generations. I mentioned I'd seen a notice announcing the recent birth of twins. She said the new arrivals would help keep the primary school going, which supported two teachers and two teaching assistants. A couple of other locals taught on St Mary's. Trish herself worked part-time on St Mary's, in adult education, giving careers advice and helping with various projects sponsored by the European Social Fund. I asked her if the weather often disrupted her commute to St Mary's, and those of the two St Agnes teachers who worked there. It did sometimes, she said, but this was an accepted fact of life on the islands. She told me she was originally an O'Neill from Port Stewart in Northern Ireland. But she'd met and married a man from St Agnes, who skippered one of the local boats. Now she was Mrs Peacock.

We did not have much time on the island before our boat was due. So we bypassed Wingletang Down, the peninsula at the southern end of St Agnes, in favour of the pursuit of the Troy Town Maze. So we missed St Warna's Well, a stone-lined spring dedicated to an otherwise unknown Irish saint of indeterminate gender, said to be the patron saint of shipwrecks. According to an eighteenth-century book on the Scillies (Robert Heath, *A Natural and Historical Account of the Islands of Scilly*, 1750), the well was a focus of ritual activities by the islanders, who believed that the saint, also called Awana, had the power to guide ships laden with valuable cargoes onto their shores. A later account describes the islanders throwing bent pins into the well to bring about such wrecks. Wells such as this were sacred to the Celts long before Christianity came to these islands. Some are much older even than the Celts. St Warna's is the only ritual well to survive on the Scillies.

Troy Town, a farm with a small collection of houses, is the most southerly and most westerly settlement in England. The eponymous maze is a few hundred yards to the south. The name Troy is associated in various places in England with mazes and labyrinths: for example, there is a turf maze in the Howardian Hills of Yorkshire called the City of Troy; and at Somerton in Oxfordshire there is a turf maze at Troy Farm. Similar mazes are found across Scandinavia, and have names such as *Trojaburg* and *Trojienborg*. Since Classical times, the legendary city of Troy was associated with mazes, it being thought that its walls were built as a labyrinth, to perplex any attackers who penetrated the outer defences. Troy also played a role in some of the medieval foundation myths of Britain, which was supposedly founded by a Trojan prince called Brutus. Brutus became the first of a line of British kings, the greatest of whom was Arthur.

We faced a confusion of crazed granite outcrops dotted along the southwestern shore of St Agnes in our search for the Troy Town Maze. Eventually we found what we were looking for, on a sloping sward looking westward across Smith Sound to the rocks of Hellweathers and Old Woman's House. The maze, set in the middle of a prehistoric field system, is ovoid in shape, perhaps a dozen yards in diameter, and consists of near-concentric circles marked out by sea-worn cobbles and turf ridges. The path takes the visitor from the entrance to either side, then inwards, then outward again before finally reaching the centre. If it had been ancient, the shape of the maze would have been buried deep beneath the surface. Someone was maintaining its shape, carving out the gravelled pathways between the ridges. Perhaps it was the feet of visitors that did the work. I was glad there was no interpretation board to explain away the maze's mystery. Later, I read that it may have been built in 1729 by one Amor Clarke, keeper of the St Agnes lighthouse (built in 1680 and still dominating the island's skyline, although now long disused).

Some years before my visit to St Agnes, as I'd walked the line of the Scottish border from the Solway to the North Sea, I had passed near to Troy Town's antipode, one of the most northeasterly places in England, north of Berwick-upon-Tweed: a farm called Conundrum. Conundrum was not always in England: in the Middle Ages Berwick and the land around it changed hands between Scotland and England thirteen times. Similarly, it is difficult to think of Troy Town, or St Agnes, or indeed any of the Isles of Scilly as being true parts of England. The islands are so far beyond Cornwall, and Cornwall itself is a Celtic realm beyond England, only attached to the rest of Britain by accidents of geography and the quirks of history. Both Northumberland and Cornwall have their own flags, often flown in preference to the Union Jack or the Cross of St George.

We had to walk fast down the lanes from Troy Town to the quay to catch our boat. As we sailed back to Tresco in the late October afternoon, the flickering waves on the top of the swell of the Road reflected a bronze and silver sky. I looked towards where the sun was heading, beyond the jagged silhouettes of the Western Rocks, beyond even the distant Bishop Light. I couldn't tear my eyes away. Something was fixing me to that far horizon. I didn't know what I was staring at, what I was yearning for. Then I realised that I was looking towards where the day was going, far to the west across the Atlantic. And I understood how the ancients conceived of, and longed for, a land over the horizon where, when they were cloaked in darkness, the sun still shone.

Never Say R*bb*t

The Isle of Portland

> The peninsula carved by Time out of a single stone . . . has
> been for centuries immemorial the home of a curious and
> well-nigh distinct people, cherishing strange beliefs and
> singular customs . . .
>
> – Thomas Hardy, 1912 Preface to *The Well-Beloved* (1892)

The Isle of Portland must have presented a grim prospect to the
first convicts shipped there from Weymouth in 1848. It was 24
November, and nothing but winter and years of forced labour lay
ahead.

From the Weymouth shore, Portland looms high above the sea,
a citadel of solid rock. Peering more closely through the gloom,
the sixty-four shackled prisoners, guarded by policemen and a
detachment of the 23rd Fusiliers, would have seen that this solidity was chipped and riddled with quarries.

The convicts knew they were being sent to Portland to break
stone – stone needed for the breakwaters that would create the vast,
fortified naval anchorage of Portland Harbour, one of the biggest
engineering undertakings of the Victorian era.

Their prison – HM Prison Portland – was to be at the Grove,
on top of the eastern cliffs. The very highest point of Portland,
the Verne, towering more than 400 feet above the sea, was to be a
huge artillery fort, surrounded on its seaward side by cliffs, and

on its landward side by a dry moat, 70 feet deep. The spoil from digging the moat was to be added to the breakwaters.

The convicts were told they were lucky. At least it wasn't New South Wales, the wardens said, or the gallows at Newgate. But conditions were harsh, and many sent to Portland over the succeeding decades did not live out their sentences. At one point the rate of deaths approached one per week. Besides those killed in quarrying accidents, and the many who committed suicide, three convicts were hanged for murdering their warders, and one died after his back took thirty-six lashes, the maximum allowed. The man's crime was to have burnt some paper and clothes in his cell. It can be cold up on the Grove.

Those who did survive their time on Portland found that they were to be sent to Australia after all. If they'd behaved themselves, they were told, they might be allowed to seek work in the colonies.

On 29 July 1849, Prince Albert visited to lay the foundation stone of the breakwaters. He presented the prisoners with a bible, 'in the hope of their amendment'. (In contrast, when Albert's son visited the prison in 1902 as King Edward VII, he ordered that each prisoner be served 'half a pound of roly-poly and two ounces of golden syrup'.) By 1851, there were over 800 convicts working in the quarries and on the breakwaters. In that year alone, a third of a million tons of rock were broken out of the cliffs. Tourists flocked from all over the country to see the convicts at work.

~

Although technically a peninsula, Portland is as good as an island. Its only natural link to the mainland is the long, narrow shingle spit of Chesil Beach, which starts at the northwestern end of Portland and runs parallel to the coast of Dorset for ten miles or more, before making landfall at Abbotsbury. It is separated from the mainland by a slender lagoon, the Fleet. At its southeastern end the Fleet enters Weymouth Bay through Small Mouth, the narrow channel of the sea that separates Portland from the mainland. Small Mouth was the moat that for very many centuries maintained Portland as, in effect, an island. Long before this, Portland *was* an island, until the sea rolled the great gravel bank of Chesil Beach slowly coastward.

The unstable shingle of Chesil Beach makes it impractical to traverse its length. Even today there is no road, not even a path, anywhere along it. Until the first bridge was opened across Small Mouth in 1839, the only practical way onto the Isle of Portland was via the ferry – a small boat pulled across the gap along a fixed rope. Not everyone was brave enough to trust the ferry. When in 1766 Thomas Newton, Bishop of Bristol, was due to consecrate the new church of St George on the island, he preferred to be carried across the waters at low tide by some sturdy locals. Thirty years later, a nervous visitor was reassured that no one had drowned since a new ferryman had taken over. The reassurance was diminished when it became known that the new ferryman had only commenced his duties nine months previously. The ferry itself was swept away in the Great Gale of November 1824. That same storm surge destroyed much of the nearby fishing village of Chiswell, at the Portland end of Chesil Beach. Twenty-six residents died, either drowned or buried beneath their collapsed cottages. It made the headlines across England: 'A tempest heavy with more frightful terrors is scarcely within the memory of man,' intoned the *Bristol Gazette*.

~

Thomas Hardy used the Isle of Portland – this 'peninsula carved by Time out of a single rock' – as the setting for his late novel, *The Well-Beloved*, in which he describes Portland as the head of a bird stretching out into the English Channel. The 'single rock' consists of oolitic limestone, formed in the shallow, warm, subtropical seas of the Jurassic Period, between 200 million and 145 million years ago. The limestone strata lie on top of sloping beds of sand and clay, an unsure foundation, leading to many landslips and rockfalls. The pale Portland stone, being both weather-resistant and easily worked, has been widely used as building material in prestige projects, from the UN Headquarters in New York and the Cenotaph in Whitehall to St Paul's Cathedral. (Christopher Wren had a long-running dispute with the Portland quarry owners, telling them 'You must not think that your insolence will always be borne with.') . Hardy describes how the Isle 'stood dazzlingly unique and white against the tinted sea', as 'the sun flashed on infinitely stratified walls of oolite'. He then quotes from Shelley's *Prometheus Unbound*, invoking 'The melancholy ruins / Of cancelled cycles.'

I think of Portland not so much as Hardy's bird but as one of Shelley's melancholy ruins – a plesiosaur. Its neck would be the long shingle spit of Chesil Beach, its tail the tip of Portland Bill, Portland's southernmost point. Such creatures swum in the seas at the time the rock of Portland was forming. One species of plesiosaur whose fossil remains were found here has been named *Colymbosaurus portlandicus*. It dates from the late Jurassic.

People came to Portland much later than the plesiosaurs. The earliest evidence of human habitation is round Culver Well, towards the south end of the peninsula. Here Mesolithic people – probably no more than twenty individuals – established a settlement around 7,500–8,500 years ago, leaving behind a midden of seashells, indicating that marine molluscs formed a major part of their diet. Settlement continued through the Bronze and Iron Ages, but most of the barrows and standing stones of these forgotten people were destroyed as building and quarrying expanded in the nineteenth century. A so-called 'Druids' Temple' was obliterated in the late 1840s when the first prison was being built at the Grove. The Romans were here too, calling the place Vindilia or Vindelis (there is still a street called Vindelis Way in the village of Fortuneswell). There are few visible Roman remains, although in 1851 convicts uncovered numerous Roman graves containing crouching skeletons.

It was on the Isle of Portland that the Vikings inflicted their first recorded violence against the English, four years before the infamous sack of Lindisfarne, often taken as the start of the

'Viking Age'. In the year 789 three ships from Hardanger Fjord in western Norway landed at Portland Bill. The reeve (senior royal official) of the king of Wessex attempted to levy taxes on them, but they turned on him and killed him. In the end the locals saw the invaders off, perhaps deploying the weapon for which Portland became famous: the slingshot. Caches of thousands of round stones have been found around the island, and in the early sixteenth century the antiquary John Leland wrote that 'The people be good there in flinging of stones, and use it for defence of the isle.' Hardy himself called Portland 'the Home of the Slingers'. However, the islanders' skill with deadly slingshots did not prevent further Viking raids over the following two centuries. As recently as the nineteenth century Portland parents would warn their children to behave, lest 'cruel wild men who come over the beach in the middle of the night' carry them off.

After the Norman Conquest, recognising the agricultural wealth of the Isle, William the Conqueror took the Manor of Portland for himself, and parts of the Tower of London were built from Portland stone. Portland had already been a royal manor in Saxon times, and William maintained this status, which meant that the islanders were not vassals of a local overlord, but paid 'quit rent' direct to the king, so owing no service, whether military or otherwise. They were represented in a 'court leet', which governed all local affairs, and which upheld the rights and privileges of the islanders. This may partly account for the fiercely independent spirit of the inhabitants, who, it is said, regard even people from Weymouth as foreigners. The court leet – properly 'the Court Leet of the Island and Royal Manor of Portland' – still exists, along with its bailiff, chief constable and reeve.

The French raided Portland on more than one occasion during the Hundred Years War, burning St Andrew's Church and carrying away sheep and cattle. Some of the sea fights of the Armada and the Anglo-Dutch Wars took place off Portland. In the English Civil War the islanders came out strongly on the Royalist side. Weymouth, in contrast, came out for Parliament. Parliamentary forces took Portland, but the Portland Royalists (disguised as Parliamentary troops) recaptured it and even (briefly) took Weymouth. Portland itself, blockaded by Parliamentary ships, was eventually obliged to surrender, albeit on generous terms.

Some might say it was typical of the Portlanders to march to a different drum than their mainland neighbours. Should any of the latter make an appearance on the island they are, to this day, referred to as 'kimberlins'. This term is more generally applied on Portland to any outsider or incomer, although coming from Weymouth is said to be the greatest crime. (A climbing friend of mine, who has explored Portland's cliffs, told me her partner was once refused a drink in a Portland pub. 'We only serve locals here,' he was told.)

Just as living on an island can encourage a sense of superiority among the inhabitants, that same isolation can also bring about the evolution of distinct dialects. At one time Portland had many words used nowhere else, not even in the rest of Dorset; it has been said that in the past the Portland dialect would have been understood by early medieval Saxons. Examples of local dialect words include 'snalter' (the bird known elsewhere as a wheatear, and formerly eaten as a delicacy on the island), 'quiddle' (squid) and 'gaberty men' (the Customs authorities). A 'weare' is a sloping pasture between the cliffs and the sea (the island has both a West Weare and an East Weare), where once there grazed thousands of Portland sheep, a breed valued both for its fine wool and for its 'exceedingly delicate' meat; but quarrying eventually ate up the pastures, and the weares are now largely choked with rubble. An 'ope' is an opening leading down to the seaside; the village of Chiswell has alleys named 'Big Ope' and, more dispiritingly, 'No Ope'.

Similarly, in the past on Portland there were a range of local customs, unknown elsewhere. For example, the seventh Wednesday after Christmas was 'Binding Day', when people would kidnap whatever of their neighbours' possessions they could lay their hands on, and would only return them on payment of a small ransom. Another custom (formerly also common in other parts of the country, but perhaps persisting on Portland longer) was that a couple would not get married until the woman became pregnant. Outsiders who abused this custom and failed to marry the woman on whom they'd fathered a child would be stoned off the island.

One custom still persists, after a fashion. 'There's a certain word you should never say on Portland,' I was told by my London neighbour, Richard Green, a long-term Portland fan. 'You mustn't *ever* say the word "rabbit".'

For thirty years Richard and his wife Jane have had a small cottage in Fortuneswell (as Chiswell is known as it creeps up the hillside, and Underhill turns into Tophill – toponymy on Portland is a little confusing to the uninitiated). I'd noticed Richard and Jane's car bore a sticker saying 'Keep Portland Weird'. Although this slogan originally emanated from Portland, Oregon, Richard assured me that the strangeness of Portland, Dorset, was a thing to be treasured and preserved. Hardy himself writes that Portland is 'the home of a curious and distinctive people cherishing strange beliefs and customs'.

Apparently the taboo on saying 'rabbit' on Portland (analogous to the ban on the word 'pig' on Lindisfarne) is recorded no earlier than the 1920s, and is said to derive from a superstition of the Portland quarrymen, who purportedly noticed that rabbits would emerge from their burrows shortly before a life-threatening rockfall. The animals were thus held to be portents of disaster, and if one appeared, the quarrymen would lay down tools until the danger was deemed to have passed. Even the former mayor of Portland, Stuart Morris, author of the invaluable *Portland: An Illustrated History*, observes the taboo, referring instead to 'a little furry creature with long ears', 'conies' and 'underground mutton'. How much of Morris's tongue is in his cheek, it is difficult to say. In October 2005, when the Wallace and Gromit film *The Curse of the Were-Rabbit* was launched, posters destined for the Isle of Portland omitted the word

'Rabbit' from the title, and employed instead the slogan 'Something bunny is going on.'

It is of course possible that extensive burrowing by rabbits (introduced to Portland as a source of food and fur in the twelfth century) might indeed cause landslips, but my neighbour Richard is not convinced. He told me of an alternative theory, that up until the seventeenth century, Portland – like many other stretches of the southwest coast – was raided by slavers from the Barbary Coast. So 'Rabat!' (referring to the present-day capital of Morocco) was shouted as a warning. 'A bit fanciful, perhaps,' Richard concedes.

~

It was a gloomy Monday in early February when we – my wife Sally and our friends Alice and Tom – drove along the causeway spanning Small Mouth to the Isle of Portland. It was not quite as Hardy describes it: 'The towering rock, the houses above houses, one man's doorstep rising behind his neighbour's chimney, the gardens hung up by one edge to the sky, the vegetables growing on apparently almost vertical planes, the unity of the whole island as a solid and single block of limestone four miles long . . .'

It was not like that because we could not see a thing. The air was thick with drizzle, and the Isle was shrouded, unseen, in low cloud. Of the sun flashing on dazzling white walls of oolite there was no sign. I thought wryly of the 1930s slogan promoting Portland as a holiday destination: 'The Sunniest Isle for Many a Mile'.

Richard and Jane's small fisherman's cottage, where they'd invited us to stay, sits in a stone terrace climbing a steep road leading from Chiswell up to Fortuneswell. At the back, as Hardy had promised, rows of cottages rose steeply above us into the mist, and a small walled garden looked over the sea – or would have if we hadn't been in cloud. But we could hear the waves flopping onto Chesil Beach somewhere below us, then the rattle of pebbles and the shingle sighing as the water sluiced back down.

This end of Chesil Beach, where it curves into the Isle of Portland, embraces Chesil Cove. This was where the local fishermen would launch their boats – sturdy clinker-built vessels known as lerrets, up to twenty feet long, with a stern as sharp as the prow, and powered by four or six oarsmen. The lerret was unique to Portland, suited to the harsh conditions of Chesil Beach, and able, according to an 1849 report, to 'live in any surf'. They would need to. Many passing ships were driven by southwesterly gales onto the lee shore here, and Chesil Cove became known as 'Dead Man's Bay'. As on the Scillies, there is no evidence of ships being lured to shore by 'wreckers' waving lights, but the local fishermen made the most of any salvage opportunities. The nearest thing to wrecking occurred in 1311, when Portlanders cut the cable of the *St Goymelote*, which was sheltering here from a storm. The ship was driven onto the shore, where it was relieved of its cargo of wine. But with the authorities taking a dim view of such plundering, the islanders were generally less proactive. Instead, they would let nature take its course, make heroic efforts to rescue the crew, and only then set to the business of sharing out the cargo. When in the 1820s the predecessor of the RNLI offered the Portlanders their own lifeboat, they rejected it, preferring to use their lerrets, and to hold onto their 'perquisites'.

The waves that hit the shore at Chesil Cove can be huge. On our first evening, as the barman poured our pints at the Cove Inn, which sits behind the sea wall at the top of the steep slope of Chesil Beach, he told us about the great storms of January and February 2014. I remembered seeing footage of the waves hitting Portland on the TV news. In the same storms, the Paddington–Penzance line was washed away at Dawlish, beyond Exeter. On St Valentine's Day, 2014, a particularly huge wave smashed through the first-floor windows of the Cove Inn, washed down the stairs, and swept the landlady down into the cellar, knocking out one of her teeth.

It was in such storms that the Portlanders made some of their boldest rescue attempts. Once the hapless ship was ground by the

angry sea into the steep bank of pebbles, one of the locals waiting on the shore would wade down into the water with a rope tied
round his waist, held by his companions further up the shingle.
The weight of the waves breaking must have been tremendous,
and the mass of pebbles washing back down the slope almost
impossible to withstand. An eyewitness account from December
1841 describes what happened when the brig *Amyntas* was blown
ashore in a storm. One of the Portlanders on the beach plunged
into the maelstrom and managed to throw a line aboard. At that
point a great wave broke over the ship, and the six crewmen who
had sought safety clinging to the foremast were swept away:

> One loud shriek was heard from those who witnessed this
> awful sight. In an instant John Hansford rushed into the surf
> and was buried amongst the white foam. After the receding of
> the wave he was seen struggling with a man in each hand, and
> although he was unable to keep on his legs he firmly kept his
> hold of them, and was hauled up on the beach by his brave
> companions, bringing with him the two poor fellows, who
> were with difficulty restored.

It did not always end so gloriously. When in 1872 the *Royal
Adelaide*, en route to Australia, was blown broadside onto the
beach, all but nine crew and emigrants of the sixty-nine aboard
were rescued. But the hundreds of barrels of wine, brandy, rum
and schnapps washed ashore attracted great crowds. Soon,
despite the efforts of the coastguard, reinforced by soldiers from
the Verne, there were hundreds of drunken people on the beach,
shouting and fighting. In the morning the bodies of several men
were found along the strand. They had died from a combination
of hypothermia and alcoholic excess.

Leaving the memory of wrecks behind us, we took the steep
path up into the cloud to the top of West Cliff. Just above our
cottage I encountered a neat-looking man in his early sixties. We
fell to talking. I asked him whether he was a Portlander. He

wasn't. His grandfather had been in the Metropolitan Police, and had been posted to Portland ('he'd misbehaved himself a *little bit*') to help guard the naval dockyard. He'd married a barmaid at the Jolly Sailor in Castletown, the former naval base on the north side of the Verne. My acquaintance hadn't actually been born on the island, but had lived there since the age of two. Now he ran a seafood business down in Chiswell, storing live crabs and lobsters in tanks for distribution round the country.

At one point, for a reason I can't recall, he said the word 'rabbit'.

'You don't mind saying that word?' I asked.

'No, no one minds these days, apart from a few old fishermen. My lads would sometimes throw a dead rabbit onto the deck of a fishing boat. The fishermen wouldn't sail. Passed on now, a lot of them. Life goes on.'

Above the cottages the path made its way upward, sometimes by steps, sometimes by muddy grass slopes. A slip here could have unpleasant consequences. Below, under the base of the cloud, we could just make out the white lines of the waves as they growled up the shingle. At the top of the cliffs the evidence of quarrying began: a stark abandoned crane, piles of spoil, sturdy gangways built of huge unmortared blocks. More than once our path was diverted away from the edge because of recent cliff falls. At one point we found ourselves in Tout Quarry, a shallow arena hacked out of the flat plateau of Tophill. The place had been turned into a sculpture park. Looming out of the mist we spotted various animals carved out of lumps of rock: a ram, a mammoth, a bear, an octopus . . . and a rabbit.

Further along, at a junction of paths above the cliffs, there was a large pointed rock, beneath which posies of plastic flowers had been placed. Someone's nickname was spelt out in large capital letters along a metal frame. A plaque on the rock was in memory of 'a loving son, brother and father'.

There was a woman nearby, with her dog. I asked her if she was a Portlander. No, she said, she was from Weymouth, but had married a Portlander and lived here happily for fifteen years. Perhaps the animosity I'd heard about was now dying out.

I asked about the memorial. 'He jumped,' she said. Then she added: 'We get a lot of jumpers along here.'

I looked down. The limestone cliff beneath us was sheer, about fifty feet high. It rested unsteadily on steep slopes of clay and scree, dotted with vegetation. If you jumped off the top you'd probably survive, for a while at least, as the soft slope below broke your fall. But it would also break your bones, inflict terrible internal injuries. You might be conscious of the pain for some time before you died.

The bleakness of the day continued as we passed by a few more cliff-top memorials, and then the village of Weston, lurking in the mist. From what we could see, Weston consisted of a series of 1960s council blocks, three storeys high, with small windows and walls made of concrete slabs. The housing looked cold and damp, and faced into the wind. Beyond Weston you come to Southwell, a village that appeared even grimmer, from the coastal path at least, than Weston. On this side it is dominated by a complex of large grey concrete buildings, reminiscent of a Soviet corrective facility, but in fact formerly the location of the top-secret Admiralty Underwater Weapons Establishment. Part of this complex is now occupied by the Isle of Portland's main school. It takes not only secondary-age children, but also infants and juniors. A few years ago it was rated 'inadequate' by Ofsted, and put into special measures. It has since rebranded itself as the Atlantic Academy.

Beyond Southwell the island tips down towards Portland Bill. This southern point of Portland features as the Snout in *Moonfleet*, a once-popular 1898 children's adventure yarn by J. Meade Falkner, largely set in this part of Dorset (the fictional village of Moonfleet is based on East Fleet, by Chesil Beach). The Bill is dominated by a distinctive lighthouse, painted white with a broad horizontal red band round its middle. Even in Roman times beacon fires were lit here to warn shipping of the dangers not only of the rocks of the Bill, but also of the Shambles, an offshore sandbank. Between the two runs the notorious Portland Race. As Trinity House themselves say on their website, 'Strong currents break the sea so fiercely that from the shore a continuous disturbance can be seen.' The Portland Race is caused when two tidal flows moving in different directions hit each other over a shallower area of seabed called the Portland Ledge. The Race can travel at speeds up to five knots. The area is very dangerous to shipping at high water, and for an hour or so afterwards. In the *Shell Channel Pilot*, the author, Tom Cunliffe, describes the sea off Portland Bill as 'the most dangerous extended area of broken water in the English Channel. Quite substantial vessels drawn into it have been known to disappear without trace.' Among the ships to have foundered on the Shambles was the *Earl of Abergavenny*, which went down in 1805. Among the 263 dead was the master, John Wordsworth, brother of the poet. The tragedy became the subject of one of William's less admirable efforts, 'The Daisy':

> Ill fated Vessel! – ghastly shock!
> – At length delivered from the rock
> The deep she hath regained;
> And through the stormy night they steer,
> Labouring for life, in hope and fear,
> Towards a safer shore – how near,
> Yet not to be attained!

The first pair of lighthouses on Portland Bill were built in the early eighteenth century, and in 1844 Trinity House erected an obelisk at the tip of the Bill, below the present lighthouse, to act as a daymark. The new lighthouse was completed in 1905, and the Old Higher Lighthouse came into the possession of Marie Stopes, the birth-control campaigner, and was for long her summer residence (she claimed her only child was conceived on the top of the tower).

At the Bill you are not far above the sea. Light was breaking through the low cloud along the southwestern horizon, and waves were racing into the rocks, the wind blowing tails of spray back from their crests. The obelisk stood firm in the gusts, unadorned apart from a plaque bearing the inscription

T H
1844

To the east along the shore I could make out, through the spray and drizzle, a series of jetties that looked like they had been carved out of the raw rock of the Bill. A simple old crane stood with its back to the wind as waves smashed and broke around it. This must have been the most unsheltered and dangerous of the many places round the island where quarried stone was loaded onto barges.

Our way home was to be an inland route, up the middle of the peninsula. The marked path to Southwell turned out to be a muddy, puddled farm track through fields of withered set-aside. Along the verges there were heaps of what could only be rabbit droppings. Richard had told me that the islanders brush aside this evidence by explaining that the local children have an aversion to brown M&Ms, and simply throw them away. In heaps.

Then I saw by the side of the path, shivering and barely moving, a dying rabbit. Its back was blistered, its eyes had already gone. Myxomatosis has cut a swathe through the Isle of Portland.

The light was failing as we reached Sweet Hill on the south side of Southwell. To avoid a long, dark walk up the road we caught the bus back to Fortuneswell. Here in the Co-Op, looking for something for our supper, I noticed that along the shelves, in anticipation of Easter, there stood rows and rows of gold-wrapped chocolate rabbits. 'Lindt Gold Bunny,' said the label. '£2.50 per 100g.'

~

A chill mist still clung around Chiswell the next morning, as we walked along the shore past No Ope. We were to explore the other side of the island, the side with the prisons. There are now two on the island: HMP The Verne, on top of the cliff-ringed hill where the Victorian citadel had been; and a Young Offenders' Institution in the original HMP Portland, at the Grove, above the eastern cliffs. There was until recently a third prison, HMP Weare, Britain's last prison ship – in the tradition of the hulks once moored in the Medway, off the Isle of Sheppey, or in Portsmouth Harbour. HMP Weare was originally built in 1979 as an accommodation barge for offshore oil and gas workers, and was later used as a prison ship by the New York City Department of Correction, docked in the East River. In 1997 it was acquired by the UK government as a solution to prison overcrowding. It was moored in the old Royal Navy dockyard on Portland, and was in use up until 2005.

Portland had for a century and a half been an important base for the Royal Navy. Portland Harbour, sheltered by its convict-built breakwaters, became home to the Channel Fleet, and in 1944 was one of the main embarkation points for the D-Day landings. In the era of the Cold War, HMS *Osprey*, the Royal Navy shore establishment on Portland, included Europe's largest and busiest helicopter base. HMS *Osprey* also encompassed the Admiralty Underwater Weapons Establishment, which was notoriously infiltrated by the so-called Portland Spy Ring. The first links in the chain were two local clerks, Harry Houghton and his lover Ethel Gee, who worked for the Admiralty at HMS *Osprey*, and who passed on material to Gordon Lonsdale, a Soviet intelligence officer whose real name was Konon Trofimovich Molody. Lonsdale used a powerful radio transmitter in the Ruislip home of Peter and Helen Kroger (aka Morris and Lona Cohen) to send the material on to Moscow. The ring was exposed in 1961. Long prison sentences were handed out, but Lonsdale and the Krogers were later exchanged for UK citizens held in the USSR. Houghton and Gee married the year after their release in 1970.

The neck of Portland, where it links to the causeway to Weymouth, is now dominated by light industry and the National Sailing Academy, host to the sailing events in the 2012 Olympics. In the past it was dominated by the artillery fort built here by Henry VIII to guard Portland Roads, an important refuge for both naval and merchant ships even before the breakwaters were built to form Portland Harbour. The enemy then were French privateers, and a twin fort was built on the Weymouth side of the bay at Sandsfoot. Portland Castle never saw action, although it was twice besieged during the Civil War, and has given a name to the surrounding settlement: Castletown.

Castletown has today something of an abandoned air. The navy has gone, and the Jolly Sailor is no longer open for business. Although one of the two massive accommodation blocks built for HMS *Osprey* has been converted into luxury flats, the other, so far unloved by developers, has been left as a skeleton of empty, open-fronted concrete boxes piled on top of each other. It would make a good stand-in for the war-shattered Holiday Inn in Beirut, *circa* 1976.

The Tourist Information Office at Castletown was far from abandoned, however. Even on this cold Tuesday in early February, it was not only open, but staffed by two helpful women. The one I spoke to had lived on the island since she was a girl. I asked what was the difference between Fortuneswell and Chiswell: 'Fortuneswell is going up the hill, Chiswell's going down,' she explained. I wasn't sure I was any the wiser. I asked about Tophill and Underhill. 'Down here we're in Underhill, whether it's Castletown or Fortuneswell or Chiswell,' she told me. 'Up there, on the Verne where the prison is, that's Tophill.'

It was a long steep climb up to Tophill. We followed a slippery path between hedges, Castletown fading below us into the mist. Near the top we encountered a bramble-covered embankment, the outer defence of the Verne Citadel. Let into the bank was a stone-lined recess. At the back of the recess was a doorless doorway. I peered inside, letting my eyes adjust to the darkness. It was a small, empty space. A few broken beer bottles littered the floor. I could just make out some faded graffiti on the back wall:

> One . . .
> makes you . . .
> small
> Go ask Alice

I recognised the fragments from a 1967 Jefferson Airplane song. The lyrics allude to *Alice in Wonderland*, and mention pills, and mushrooms, and a hookah-smoking caterpillar. The song's called 'White Rabbit'.

We followed the bank until the path turned south. Through the murk we could make out that we were in the dry moat that forms the second line of defence of the Citadel. It was hemmed in by towering stone walls. Steps led up to the outer rampart. Here the defile is spanned by a more recent concrete footbridge, blocked by a barred gate. A sign says

> SOUTH GATE
> HM Prison Service
> IRC The Verne

The military had left the Citadel after the Second World War, and in 1949 the place was established as a prison for men serving medium- and long-term sentences. IRC stands for Immigration Removal Centre, a role the Verne played between 2014 and 2017.

For those who have experienced both, detention in an IRC is worse than prison. Volunteering in a migrant support centre in London a few years ago, I met a man who had fled to the UK from the Democratic Republic of Congo because of his opposition to the regime. One of his friends had been killed, others had disappeared. His asylum claim was rejected by the Home Office. Desperate to find a way to support himself, he began to work, using a fake passport. He was arrested and sent to prison. The

English prisoners couldn't understand why he was there. 'Because you work they put you in prison?' they asked. After eight months he was released, but then he was detained again and placed in an Immigration Removal Centre. 'I was so depressed I want to kill myself. I thought detention would be better than prison. I was wrong. Because all this time no one tell you when you'll be deported or released. You don't know nothing. That was hell.' (Many years later, he put in a fresh asylum claim. This time he was successful.)

Although there are still a number of IRCs round the UK, the Verne has reverted to its role as a conventional prison. But through the mist, we could see nothing of it. A sign warned us: 'No unauthorised entry'. Another listed all the items you were not allowed to bring into a prison, and the attendant penalties should you do so.

More cheeringly, a third, less official sign pointed us towards something called 'Fancy's Farm'. We passed by high wire-mesh fences, which I took to be the perimeter of the prison. Then through the fences I saw, scattered across the grass, several dozen discarded Christmas trees. They were on their sides, the needles mostly gone. What were they doing there? Was there one tree for every Christmas spent inside by some sad lifer?

Then I saw a wallaby. I blinked. Then there was another wallaby. Perhaps, I thought, I should go ask Alice. Further on I found an explanation. There was a herd of pygmy goats browsing on the Christmas trees. And then half a dozen black-and-white piglets. Fancy's Farm was a community petting farm. But today there were no visitors.

Past Fancy's Farm, still thick in fog, we entered a labyrinth of earthen ramparts. This, the map told me, was the High Angle Battery. I'd asked Richard about this. He emailed me back: 'The construction of the battery, which was designed to pummel enemy shipping (i.e. the French) with howitzers, was started in the late nineteenth century and completed in 1912, just in time to be made obsolete by aircraft.' It was too foggy to find the entrance to any of the 'ghost tunnels', passages built into the earthen ramparts to store shells. In 2016 the artist Lee Berwick used the High Angle Battery for a strange and haunting sound installation. There is a film about it on Vimeo. All I heard when I was there was the lonely cry of a crow.

The next sign I saw was on another high wire-mesh fence. 'Please report any suspicious activity,' it said. At the top of the sign was a small logo, showing the Portland Bill lighthouse, capped with a crown, sending its cheery beams out over the waves. 'Portland HMP YOI', it said. It was the Young Offenders' Institution. Another sign further on warned that 'Security dog handlers operate in this area', while a third sign stated that 'Throwing anything into a prison establishment is a criminal offence.' Beyond the barbed-wire topped fence, I could make out a high concrete wall.

From the Grove we walked through the mist to the village of Easton, in the middle of the island. There was a shop here called Island Beauty, and another with a window display featuring toilet rolls, wet wipes and a plastic bottle of bleach. A small building on a corner called itself the Alessandria Hotel, proudly proclaiming itself 'Italy on Portland'. A narrow frontage featuring beer-swilling pigs was Shakey's, which offered 'Black Country Pork Scratchings & Bar Supplies'. We found a pub to have lunch. Except, as it was Tuesday, they weren't serving food. But they said we could buy some sandwiches from the Co-Op down the road and eat them with our pints. So we did. We fell into conversation with a young mother, whose baby had taken an interest in our whippets. Her older daughter, aged six, attended the academy in the concrete gulag in nearby Weston. When her daughter reached secondary age, she said, she'd send her to Weymouth. Another kind customer gave our whippets a packet of pork scratchings. They ate them up enthusiastically. Then sicked them up on the floor.

As you leave Easton to the south, you pass an old house with its name carved into a block of Portland Stone. It's called Rope's

End. I thought of the alley in Chiswell called No Ope, of the convicts hanged for killing their warders, of the many prisoners in HMP Portland who had tried to end their lives. We descended past the ruined cliff-top keep of Rufus Castle – named after William Rufus, son of the Conqueror, but dating from the fifteenth century – down towards Church Ope Cove. Church Ope Cove – a stretch of relatively level shingle – had been one of the favoured landing places for Portland's many smugglers.

Smuggling – along with fishing, stone quarrying and the plundering of wrecks – was long one of Portland's main industries. With its defiance of authority, smuggling appears to have suited the islanders' spirit of independent-mindedness. In 1746 a Revenue Surveyor called Warren Lisle reported that customs officials (the much-mocked 'gaberty men') were reluctant to visit Portland 'for fear of being knocked in the head by a volley of stones'. That same year Charles Wesley, the Methodist leader, preached on the island. Of one gathering he wrote, 'Some wept, but most looked quite un-awakened.' Few of the islanders, he wrote, 'as yet feel the burden of sin'. (His preaching eventually had some effect, however. 'My mouth and their hearts were opened,' he confided to his diary on 9 June 1746. 'The rocks were broken in pieces, and melted into tears on every side.')

The Portland smugglers had turned their business into a fine art. They would rendezvous with French ships out at sea, and attach the barrels of brandy to the underside of rafts. The rafts would be towed nearer the shore, where they would be temporarily sunk, the place being marked by a small buoy. Then at dead of night the smugglers would row out a lerret to retrieve the contraband, or would employ the services of a specially bred 'Portland sea dog', which would swim out and bring ashore a line attached to the smuggled goods. (These large sea dogs also saved many victims of shipwrecks.) Once ashore, the contraband would be hidden in cellars or recesses set into a cottage wall. The women of Portland played their part. They would conceal lengths of silk in their undergarments, and 'bladders' of brandy under their dresses. Smuggled goods taken for sale in Weymouth were often touted as 'freshly found on the beach'. It seems the islanders were a canny lot; as early as the sixteenth century John Leland had observed that they 'be politic enough in selling their commodities, and somewhat avaricious'. The fact that there is an alley in Chiswell still called Brandy Row indicates that Portlanders are quietly proud of their smuggling heritage.

Descending towards Church Ope Cove, our path turned north along the weare under Grove Cliff. In the mist there was no sign of the thousands of sheep that once grazed here, their presence now only remembered by the name of the stretch of ground along the top of the cliff: Shepherd's Dinner. (Only a handful of Portland sheep still remain on the island, at Fancy's Farm.) The weare is now a mix of rubble and scrub, with here and there a sewage pipe and the remains of wartime defences. On one slab of concrete, of no determined purpose, I found some enigmatic graffiti: nine identical red rabbit silhouettes had been painted in a line, all facing left. Next to them someone had stencilled the words

WE
ARE
HERE

Our path eventually climbed up to the Young Offenders' Institution at the Grove. HMP Portland had become a Borstal (a detention centre for delinquent youths) in 1921, and a YOI in 1988. The tall chimney stacks and severe edifices of the original Victorian prison still dominate this stretch of the island – or would have, if they had not been hidden in fog. During the Second World War, one block was bombed by the Luftwaffe, killing four boys; another boy was later killed after he escaped and trod on one of the many landmines laid along East Weare.

A sign told us our intended path back to Church Ope Cove, along Shepherd's Dinner, was closed. There had been a cliff collapse. So we skirted under the walls and high fences that ring the YOI, aiming to return to Easton that way. Above one stretch of wall a spotlight on a high post shone through the murk of the late afternoon. Around the corner we walked past the front gates of the YOI – one gate modern, one Victorian. In the shelter of the arch of the latter, two prison officers were having a smoke. The older man was grey haired and clean-shaven, the younger sported a large, sculpted beard. If it had not been for the uniform, he could have been a hipster from Hoxton or Hackney Wick.

I asked if I could take their photograph. 'It's against the law,' said the man with the beard. I asked whether I could take a photo of the Victorian gateway without them in the picture. 'It's illegal to photograph the outside of a prison,' I was told firmly. I later failed to find any evidence that this is the case. Identical signs outside both HMP The Verne and the YOI listed a range of offences under the Prison Act 1952. The signs stated that you could be sentenced to '2 years imprisonment or an unlimited fine or both' if, without authorisation, you took a photograph *'within a prison'* (my italics).

Just beyond the YOI, on our path back to Easton and the bus home, we passed behind a row of small terraced houses. In the garden of one there was a flagpole, from which a Cross of St George flapped in the fog-filled air. Below it, hanging limply, was a Confederate battle flag. The latter has become associated with far-right, white supremacist groups, not only in the USA, but also in Europe. I wondered what it would be like to be a black inmate of HMP Portland.

~

We needed to be back in London the next day. In the morning, I just had time to explore a little of the West Weare in some rare sunshine. Perhaps Portland was, after all, 'The Sunniest Isle for Many a Mile'.

I'd heard that the best view of the island was from a hill above Abbotsbury, at the far end of Chesil Beach. Indeed, it was said to be one of the best views in England. The sun continued to shine as we drove onto the mainland at Wyke Regis and then west through the chocolate-box villages of southern Dorset. This was a different world from the bleak concrete dilapidation of Portland, a less honest and more moneyed world where class and social status still held sway. On Portland, you are just another human being, an equal, even if you do come from off-island. Back in rural England, the England of old churches, manor houses and golden cottages, we returned to a fantasy version of the past where everyone is supposed to know their place.

I parked by the Abbotsbury Swannery, and made my way on foot up steep grass slopes towards St Catherine's Chapel, a well-preserved ruin sitting starkly on the skyline above me. It is thought to have been a medieval pilgrimage destination. The chapel survived the Dissolution of the Monasteries, as its commanding position above the sea made it an ideal place for a beacon to guide those making the difficult passage across Lyme Bay.

Walking round to the seaward side of the now empty chapel I peered along the length of Chesil Beach, stretching for miles towards the Isle of Portland. But although I stood in sun, and sun shone on the green hills to the north and west of me, the shingle spit of Chesil, extending in the opposite direction, disappeared into mist. In the far distance I could just make out, for a moment or two, a great block looming above the unseen sea. But soon the island folded its shroud around itself once more, and returned back into obscurity.

Across the Sands of Dee

Hilbre Island

'O Mary, go and call the cattle home,
And call the cattle home,
And call the cattle home
Across the sands of Dee';
The western wind was wild and dank with foam,
And all alone went she.

– Charles Kingsley, 'The Sands of Dee' (1850)

Kingsley's lines ran through my head as I contemplated the flat, damp expanse before me. It was a chill evening in late May, and in the morning I was to make my visit to Hilbre. Only a low lump broke the horizon to the northwest, where the broad estuary of the River Dee opens out into the Irish Sea. Behind me was the Wirral. To the west, beyond Salisbury Bank and the channel called Wild Road, lay Wales. I looked down at my map, once more read the notice printed in stark red letters:

> WARNING:
> It is dangerous to cross these sands
> to Hilbre Island.
> Please consult the local tide tables.

I'd checked and re-checked the tide times, consulted on the safest route, knew I must wait the prescribed three hours after

213

high water. Kingsley's next verse beat out its rhythms with a solemn inevitability.

> The western tide crept up along the sand,
> And o'er and o'er the sand,
> And round and round the sand,
> As far as eye could see.
> The rolling mist came down and hid the land:
> And never home came she.

A shiver ran up my spine.

The low lump I was looking at, a solid block of red sandstone amidst the featureless, bird-rich flats, was described by William Camden as 'the small, barren and hungry isle called Il-bree'. Hilbre Island, together with two smaller companions, Little Hilbre (or Middle Eye) and Little Eye, makes up a miniature archipelago not far from the Welsh border, which here runs down the middle of the estuary of the Dee. The islands are administratively part of the Metropolitan Borough of Wirral, and so are considered English. It appears that the three islands, separated at high tide, may have once been a single landmass; they are shown as such on John Speed's map of 1610 (not necessarily firm evidence), and a decade earlier, en route to Ireland to take on the Irish chiefs, the Earl of Essex reportedly encamped 4,000 foot and 200 horse on Hilbre – hardly possible today, when the island has an area of little more than 11 acres. This diminution and fragmentation is most likely to be due to the action of the sea over the centuries, as it constantly erodes the soft sandstone rocks of which the islands are composed. Over recent decades, the shrinkage of Little Eye has been particularly noted. Some have maintained that Hilbre was at one point in the not too distant past permanently attached to the mainland, a state of affairs held to be reflected in an old saying:

> From Birket Head [Birkenhead] to Hilbre
> A squirrel could jump from tree to tree.

Counter to this, we have John Leland's account in his *Itinerary* of 1541: 'This Hillebyri at the flood is all environed with water as an isle . . . and at the ebb a man may go over the sand.'

Hilbre (pronounced 'hilbri' or 'hilbree') is named after St Hildeburgh, who has also lent her name to the parish church of nearby Hoylake. St Hildeburgh is said to have lived on the island as a hermit in the seventh century, although some contend she never existed, while others suggest she was, in fact, St Ermenhilde, mother of St Werburgh, to whom Chester Cathedral is dedicated. In the Middle Ages, possibly in the later eleventh century, the Benedictine monks of Chester Abbey established a chapel on Hilbre, dedicated to St Hildeburgh, or to the Virgin Mary. Hilbre was then called *Hildeburgheye*, the *-eye* suffix being from either Old English *eg* or Old Norse *ey*, both meaning 'island'. (The names of Hilbre's two companions, Middle Eye and Little Eye, presumably share the same origin.) St Hildeburgh's Island became a place of pilgrimage, where intercession was sought from 'Our Lady of Hilbyri'. There is a story of such an appeal concerning 'Richard, the young Earl of Chester' (presumably Richard d'Avranches, second Earl of Chester, who died in 1120). When returning from a pilgrimage to St Winifred's Well (Holywell, in North Wales), the earl was set upon by 'Welsh marauders'. Fleeing eastward towards England and safety, he and his retinue reached the Welsh side of the estuary of the Dee. Here 'the earl addressed his prayers to St Werburgh, when the waters of the Dee were instantly parted, and an immense sandbank presented itself, over which the soldiery under the command of Fitz-Nigel, baron of Halton and constable of Chester, marched to his assistance'.[*] In an older version of the story, Fitz-Nigel and his men, anxious to rescue the earl, got as far as Hilbre, where a monk advised the constable to pray to St Werburgh. This resulted in the parting of the waters, and, as a consequence, the area to the west of Hilbre

---

[*] William Williams Mortimer, *The History of the Hundred of Wirral* (1847)

became known as 'the Constable's Sondes'. In such estuarine localities, the distinction between dry land and water is rarely fixed, and it is possible that under exceptional conditions of wind and tide, the 'miracle' described might just have taken place.

Later generations were inclined to scoff. Was the story merely concocted to boost the prestige of Chester Abbey and its patron saint, and to drum up business for the pilgrimage destination that was Hilbre Island? The Tudor Protestant propagandist Raphael Holinshed described Hilbre as 'a sort [resort] of superstitious fools, in pilgrimage to our Lady of Hilbre, by whose offerings the monks there were cherished and maintained'. In 1656, Daniel King, following Holinshed's sceptical line, cast doubt that monks would ever have set up home on Hilbre, 'for that kind of people loved warmer seats than this could ever be'. The authorities proved more pragmatic, however, opting for commerce over doctrine. After the Dissolution of the Monasteries, when Hilbre passed into the possession of the Dean and Chapter of Chester Cathedral, two monks were allowed to stay on the island to maintain a navigation beacon used by shipping. Eventually, in 1856, Hilbre was sold for £1,500 to the Trustees of Liverpool Docks (later the Mersey Docks and Harbour Board).

The fishermen of the Dee were more inclined to put their faith in the local seals, which when the fog came in were said to bark to show the fishing boats the way through the treacherous waters. Hence the fishermen refrained from killing the creatures, and were happy to let them have a share of the salmon swarming up the River Dee to spawn. Grey seals still pup on Hilbre, having survived the efforts of the Fishery Board in the 1950s to reduce their numbers by organised shooting.

In the sixteenth and seventeenth centuries, Hilbre was a bustling little port, mostly trading with Ireland, and there was even a small factory refining rock salt. At this time, there was always a lagoon between the offshore sands of Hoyle Bank and the northwest shore of the Wirral, an area of water where ships could anchor without being grounded, even at low tide – a feature reflected in the name of the settlement here: Hoylake. Hilbre probably acted as a break-water for this once busy anchorage. But silting in the mouth of the River Dee filled in the 'lake', and made the channels up to Chester increasingly unnavigable. Trade transferred more and more to ports on the River Mersey, on the other side of the Wirral.

By the time Richard Ayton visited Hilbre in 1813, the place had gone into decline, as he describes in his *Voyage Round Great Britain* (1815):

> ... there is a public house, the only habitation, and a few rabbits, the only quadrupeds, to which nature supplies a very meagre provision, only parts of the island being covered with a scanty sprinkling of grass ... The approaches to the land, between the mouths of the Dee and the Mersey, have a most formidable aspect, and a stranger casting his eye over the puzzling confusion of banks which break the sea, would scarcely believe that these dangerous passes are avenues to the great port of Liverpool ... Nothing could be more wild and dreary, and the eye was not relieved on turning to the land, which was also sand, with something of vegetation but not of verdure upon it, and without a single tree.

Ayton reports that the only permanent inhabitants were the inn-keeper and his wife, who made a little money from 'the crews of some small vessels which find a harbour under one side of the island'. But, he adds, 'their riches have been gained principally by wrecking, for which business their situation here is said to be admirably calculated'. As elsewhere, 'wrecking' may have just involved salvage, rather than deliberately luring ships to their doom by showing false lights.* Nature here provides sufficient

---

* It wasn't all just harmless *Whisky Galore*-type fun. The Royal Commission on establishing a police force reported in 1837 that 'On the Cheshire coast not far from Liverpool, they will rob those who have escaped the perils of the sea and come safe on shore, and mutilate dead bodies for the sake of rings and personal ornaments.'

hazards for shipping without requiring the aid of human duplicity.

False lights *were* shown hereabouts during the Second World War, however. In order to convince the Luftwaffe that the Dee was the Mersey, electric lights were placed on poles in the sands round Hilbre, and on Middle Eye diesel generators belched smoke to make it look like Liverpool was burning. Today, although there is a small unmanned lighthouse (built in 1927), a bird observatory and a handful of holiday houses on Hilbre, there are no permanent inhabitants. When Wirral Council advertised for a resident warden in 2011, they found no suitable candidate who was prepared to live without mains electricity or running water.

~

I confess that I'd barely heard of Hilbre Island, but, like many others of my generation, as a child I'd been read 'The Sands of Dee'. The poem was first published in Kingsley's novel *Alton Locke* in 1850, but soon acquired a life of its own. Kingsley himself based the poem on a local story of a drowned girl, who probably met her fate in the salt marshes a little further up the estuary, by Neston. He might also have heard of the fate of the wife of a warden on Hilbre, who had taken up his post in 1842. One winter, the warden's wife was returning to the island in her pony trap when she was caught by a snowstorm, lost her way, and drowned.

Kingsley shares that maudlin Victorian fascination with the death of young women that manifests itself in innumerable girl-angels adorning the necropolises of England, and, notably, in John Millais's near-contemporary painting, *Ophelia*. Millais's painting depicts the eponymous young woman maddened by grief as she floats in a stream before she drowns. Shakespeare, through the mouth of Gertrude, aestheticises the calamity, telling us how Ophelia sings before she is pulled to 'muddy death':

> There is a willow that grows aslant the brook
> That shows his hoar leaves in the glassy stream . . .

Millais follows suit in turning the death of a young woman into an object of beauty, paying meticulous attention to the flowers and weeds with which the drowning figure is surrounded. (Millais paid less attention to the welfare of his model, Elizabeth Siddal, whom he had lie fully clothed in a bath through the winter of 1851–2; she caught a severe cold, possibly pneumonia, and her father demanded £50 from Millais for medical expenses.) Kingsley's victim, though a nameless cowherd rather than the daughter of a lord chamberlain, is similarly apotheosised in death:

> 'Oh! is it weed, or fish, or floating hair –
> A tress of golden hair,
> A drownèd maiden's hair
> Above the nets at sea?
> Was never salmon yet that shone so fair
> Among the stakes on Dee.'
>
> They rowed her in across the rolling foam,
> The cruel crawling foam,
> The cruel hungry foam,
> To her grave beside the sea:
> But still the boatmen hear her call the cattle home
> Across the sands of Dee.

I had been warned. So I not only worked out the safest time and the safest route to cross the sands, I also packed my compass and plenty of warm clothes against the fog and chill that might descend, should the weather turn. I did not want to be mistaken for weed or fish or drownèd maiden's hair, I did not want to be lost in the rolling mist, caught by the western tide. When I messaged my host, Chris Davidson, at the Airbnb I'd booked in West Kirby that I hoped to avoid sharing the fate of Kingsley's heroine, he messaged back, 'We'll look after you, don't worry :-)'

~

'We're at the neap at the moment,' Chris told me. It was a wet afternoon, and he'd picked me up from the station in his car. 'You won't have to wait three hours, just follow the tide out.' This seemed rather bold to me, hubristic even. Would it be wise to defy the cruel crawling foam, the cruel hungry foam, in such a presumptuous manner?

Caution kept with me in the morning as I walked along the causeway on the seaward side of West Kirby's artificial Marine Lake, a large area of water where people can sail and windsurf at any point of the tide. Thick clouds hung low in the sky, and sharp showers gusted across the estuary from Wales. I pulled on as much clothing as I had with me. The brown, choppy waters of the Dee were ebbing. A pair of shelduck – goose-sized, harlequin-patterned seafowl – picked at the worm casts uncovered by the receding tide. Beyond them, channels of seawater snaked through the flats towards Hilbre.

The walk across to the island starts at the Dee Lane Slipway at the north end of the Marine Lake. There is an RNLI lifeguard station here. Its flags snapped in the breeze, and nearby were parked a quadbike, a surfboard and an inshore rescue boat. I asked one of the young lifeguards on duty whether they were often involved in rescue dramas. 'Not really,' he said. 'We mostly just offer first aid to people who've slipped on the rocks.' As it was still too soon after high tide to cross over, I wandered into West Kirby for a bite to eat. I found a café crammed with respectable-looking persons of a certain age enjoying cake and morning coffee. West Kirby, together with its neighbour Hoylake (with which it shares the Royal Liverpool Golf Club), is perhaps the most genteel part of the Wirral, a peninsula that sets itself apart from its more rowdy neighbour on the other side of the Mersey. West Kirby was clearly a place to retire to. It is packed with late Victorian and Edwardian villas, stone-built or half-timbered, and edged with quiet avenues of more modern bungalows.

By the time I returned to the Dee Lane Slipway, the prospects for the day had been transformed. Not only was the sky brightening, but in the distance, far out across the still-wet sands, figures were scampering along the low skyline of Little Eye. I rubbed my eyes. Ahead of me, between the mainland and Little Eye, dozens of people, many clustered in family groups, were making their way towards the islands. It was half term, and it rapidly became clear that a walk out to Hilbre at low tide was a popular day out with the kids. So much for my imagining myself all alone with the elements, braving the dangers of tide and foam and fog.

I later learnt that Hilbre had been a magnet for day trippers since at least the 1880s. There was for some years controversy over public access. The Mersey Docks and Harbour Board were ambivalent about visitors. Initially, it conceded, these were of the genteel sort, but the building of the railway to West Kirby brought less desirable types. By 1910 the Dock Board was complaining to Hoylake Council (who had requested the owners to install a public convenience) that 'a very nice class of people . . . have been displaced by a very undesirable class of roughs, both male and female . . . taking with them intoxicating liquor and playing cards'. August Bank Holiday 1911 saw 2,000 visitors come to the island. A ticketing system was introduced. The warden would take the names and addresses of any visitors who did not possess a ticket, and subsequently the offenders would receive a solicitor's letter.

In time, the attitude of the Dock Board to visitors relaxed. A handful of tenancies were offered, and a few cottages built by people who wanted to enjoy the island for more than a day. I was told that two families had owned cottages here for more than a hundred years – nowadays, the cottages cannot be sold, only passed on within families. I was also told that a few years ago one old man, not trusting that his children would respect the island, had his house demolished rather than leaving it to them in his will.

One of my informants, Charles Warren, has had a long connection with Hilbre, partly through his role as captain of Mersey

Canoe Club, which occupies the smartest building on the island, a green-painted wooden bungalow. The club was founded in 1871, and still maintains a small fleet of Victorian clinker-built sailing canoes. Charles told me that in the nineteenth century, four or five sailing clubs had bases on the island. For him and his friends, the mainland is 'the other island'.

Charles filled me in on some other aspects of Hilbre's history, and shared with me his copy of a now out-of-print book, *Hilbre: The Cheshire Island* (1982, edited by J. D. Craggs). The old inn on the island, known as the Seagull, being isolated from the attention of the forces of law and order, was the venue for various illegal activities, particularly cock-fighting and bare-knuckle boxing (activities recorded on some other English islands, such as Canvey). After the Dock Board acquired Hilbre, it built a station here as one link in its visual telegraph system connecting Holyhead in Anglesey with Liverpool. In clear weather, semaphore messages regarding shipping could be sent with extraordinary rapidity: in 1830, a signal from Liverpool reached Holyhead, and a reply was sent back and received in a time of twenty-three seconds. There is a story of one Holyhead telegraphist, irritated at being asked to repeat a signal, sending the message to Liverpool, 'You are stupid.' The reply came back: 'You are dismissed.' The visual semaphore system was superseded on this route by electric telegraph in 1861, but the station on Hilbre was kept on, and re-equipped with the new apparatus. The telegraph building on Hilbre still stands; there is an outside staircase up to its roof, and a semicircular glass-fronted room facing out to sea. Less well preserved is the old lifeboat station at the north end of the island. It was built by the Dock Board in 1849, and used when tide conditions at Hoylake made it impossible to launch from there. When there was a call-out, the crew would make their way across the sands to Hilbre by horse and cart, and launch a boat from the island. The RNLI took over the station in 1894, and closed it down in 1939, when the Hoylake station acquired a tractor to take its boat across the sands at low tide. The Hilbre station was involved in no notably dramatic rescues, but was nevertheless responsible for saving twenty-one lives. The red sandstone walls of the station still stand, but the roof has gone.

~

I followed the holiday crowds out across the drying flats to Little Eye. In some places the sand was still puddled with seawater, in some places it was muddy, in other places gritty with finely shattered seashells. The casts of marine worms were everywhere, and in many places the sand was rippled with the pattern of the retreating waves. Along the horizon, far out in Liverpool Bay, stood rows and rows of wind turbines.

It takes less than half an hour to walk out to Little Eye, known by locals as 'England's smallest island', the definition of an island hereabouts being an area (sometimes) surrounded by water, and large enough to graze a single sheep. There are a few square yards of turf on Little Eye, but it mostly comprises pancakes of red sandstone piled on top of each other. I passed an elderly couple walking with sticks. On a piece of ruined masonry of unknown purpose someone had scraped DEATH ROCK.

North of Little Eye the safe way takes you along the inland side of a stretch of slippery rocks to Middle Eye. The way was marked by numerous footprints in the soft sand. A red-uniformed lifeguard riding a quadbike drove past; the shepherd was keeping an eye on his flock. Middle Eye is not much larger than Little Eye – perhaps two sheep could graze here on the greensward, amidst the abundant pink of thrift and the yellow and red of bird's foot trefoil.

It's only a short walk across flat rocks from Middle Eye to Hilbre itself, which, in the nineteenth century had just enough grazing for the keeper's single cow, a horse and a handful of sheep. Most people take the path north along the spine of the island, but I decided to explore the west coast. At first, ropes and notices warned me away from the crumbling cliff edge, but further along the ground became more sound. Low walls separated the rocks at the top of the cliffs from the turf. I later learnt that these mysterious structures had been built in the 1970s by volunteer schoolchildren to prevent erosion of the rock and turf by wind and wave. When it seemed safe to do so, I made my way cautiously down to the shore, careful not to slip on the rocks, smeared as they were with livid green algae. If I fell here and twisted my ankle or banged my head, I might become prey in a few hours to the incoming tide. 'Oh! is it weed, or fish, or floating hair?' the finders of my body might ask. I did not want to lie, as Kingsley's heroine lay, in a grave beside the sea, so I paid heed to every step.

Horizontal strata of bright red sandstone twisted along like thick rope cables above the seaweed-strewn shore. Sometimes the cables mutated into layers and layers of overlapping slabs of flesh, then twisted into flayed bands of tendon and muscle, before sprawling out like some giant jellyfish, pulsing to a beat too slow for humans to count. Towards the north end of the island the cliffs rose in height, perhaps to twenty feet. Half that height comprised well-dressed and mortared masonry. This had been built in the later nineteenth century by the Dock Board to prevent cliff erosion and the loss of vital infrastructure on the island.

Crunching along over deep piles of mussel shells, I turned a corner and found a small sea cave set into the masonry. It had the tangy, off-fish smell of a shore at low tide. Could this be Lady's Cave? I'd heard there was a cave with this name on the west side of the island. The story behind the name, like Kingsley's poem, involves the death of a young woman. At some point in the Middle Ages, we are told, the daughter of the governor of Shotwick Castle, not far from Chester, fell in love with a young man of whom her father disapproved. Determined to have his way, the governor sailed with his daughter across the Dee to Wales where she was to marry a man of his choice. During the voyage, the father told the young woman that he had had her lover killed. Hearing this, she threw herself into the sea. She was eventually washed up on Hilbre, where a monk found her dying in her eponymous cave.

Many centuries later another unsettling event involving a young woman was reported from Lady's Cave (as subsequently recounted in the *Wirral Globe*). One winter's day in 1954, a thirteen-year-old girl called Susan Rogers had walked out to Hilbre with her eighteen-year-old cousin Tina Jones. The two apparently had a row, and Susan ran off and hid in Lady's Cave, ignoring her cousin's shouts that the tide was coming in. After a while, in the semi-dark of the cave, Susan heard a rattling close by. Then something brushed against her ankle. She looked down. A thing looking like a cane covered with bristles was quivering between her feet. It then began to prod at her skirt. She looked up. The 'cane' turned out to be one of the antennae of a monstrous, many-legged crustacean, some four feet high and six feet wide. It had blood-red eyes, and its mouth opened and closed with a rattling sound as it moved towards her. Susan was too terrified to scream. Instead, she leapt out of the cave, fell over on the rocks and sprained her ankle. Crawling along the beach, she almost passed out with fear, hearing the rattling somewhere behind her. Eventually, Tina found her in a pitiful state. The two somehow made their way off the island, to safety. Reading this story, I wondered what Dr Freud would make of it.

I mentioned this tale, somewhat tentatively, to Charles Warren. He'd never heard it. I detected a certain scepticism. What was more, he told me the little cave I'd visited and described to him was not Lady's Cave. That was further south, in the area roped off because of the unstable cliffs. Charles told me that recently a rockfall near the cave had exposed some ancient dinosaur footprints.

Leaving the small cave that wasn't Lady's Cave, I climbed up some steps towards the plateau. The wind rose. I sheltered with some other visitors by the old lifeboat station. Far to the west, along the North Wales coast, the rugged peninsula of the Great Orme jutted out into the Irish Sea. To the left of it, rising high into the clouds, I could just make out the rounded shoulders of the distant Carneddau, in Snowdonia. Someone told me that on a clear day, looking north, you could see the Blackpool Tower – even the hills of the Lake District. I could only see as far as the docks at Bootle.

In the waters below the lifeboat station a seal's head bobbed briefly in the water, and then was gone. That's all I saw of the local seals. I'd been told that they often haul themselves up in large numbers on the sands at the edge of the Swash, the channel that runs from the mouth of the Dee along the north shore of the Wirral, outside Hoyle Bank. But not the day I visited.

By now the sun had come out. I walked south to the cluster of empty buildings on the lee side of the island. Day-trippers popped in and out of the composting toilets, clambered up and down the rickety steps on the outside of the old telegraph station, sat and ate their picnics. Children shouted and laughed and played. It was, for them, the day they had their very own island adventure.

I wandered away, looking for a quiet spot. Just above the picket fence ringing the green hut of the Mersey Canoe Club, I found a patch of soft, deep grass. I sat, then lay down, out of the wind. The children's voices came and went, then finally receded. In the quiet warmth, I nodded off, happy to surrender to what the afternoon might bring. When I woke, the people had all gone. I panicked for a moment, looked at my watch. Still plenty of time before the tide came in. A martin swooped low. A rock pipit looped through the air, crying *tsip tsip tsip*, and then dropped down to the eastern shore.

Hilbre is above all else a place of birds, a staging post for migrants bound both north and south. For many years, after the lifeboat station and the telegraph station closed, the warden of the bird observatory, established in 1957, was the only permanent resident on the island, sometimes accompanied by their partner. The crime writer Ann Cleeves spent some years here with her husband Tim, who was appointed warden in 1977. Her account of the trials and tribulations of their first year forms the third chapter of *Hilbre: The Cheshire Island*. As well as keeping an eye on the island's birds, the couple also acted as coastguards, raising a canvas cone up a mast as a gale warning to shipping, and maintaining the tide gauge at the north end of the island for the Dock Board. They had to struggle with a temperamental generator, a complex system for collecting rainwater, and the difficulties of crossing to the mainland in poor conditions. On one occasion, in a thick fog at the end of October, she found herself walking round in circles on the sands following her own footsteps. She only managed to locate the dry land of the Wirral once the fog lifted, otherwise she would have found herself drowned in the Swash. A month later, crossing from the mainland head-on into a gale, the wind was so strong she could barely breathe. Reaching Little Eye, she was blown clean off her feet. This wild westerly was presaged by an influx of kittiwakes, and when the storm came it brought with it flights of petrels and skuas. Cleeves delights in listing the many seabirds and shorebirds that passed through, among them eider, purple sandpiper, black guillemot, velvet scoter, various divers and terns, dunlin, scaup, sanderling, whimbrel, gannet, and all kinds of gulls, some rarities – Iceland, Mediterranean, glaucous, laughing.

In summer the birds were fewer, but the season had its consolations: 'On several warm evenings we saw ghost moths in the garden, with silky white upper wings and dark charcoal under wings. These moved up and down horizontally about three or four feet from the ground, as if pulled on invisible strings, just as dusk began to fall.'

September brought a return of migrant activity, including a spectacular passage of Leach's storm petrel in the second half of the month. Thousands of these small, fork-tailed, open-ocean birds had been funnelled from their southerly migration into Liverpool Bay by strong westerlies. Impelled by instinct and who knows what inner map or compass, they took advantage of calm spells to fly past Hilbre out to sea in a bid to resume their southward journey.

It was time for me too to resume my journey. My landward route took me along the foot of the low, greasy cliffs of Hilbre's eastern shore. The burnt gold of wild wallflowers clustered in cracks, white cushions of sea campion tumbled over the rim. Lower down, the meagre stems and flowers of scurvy grass clung to the bare rock. Also scraping a living on the sandstone was a plant I'd never seen before. It had creeping stems and short, succulent, salt-resistant leaves, and the flower itself was a small but striking five-petalled star, deep pink, with bright yellow pistils and stamens in the centre. I later identified it as some kind of sand spurrey. Here and there, the cracked pipes of the island's disused sewage system fed down from the cottages above. Although I was in the lee of the cliffs, I could hear the wind rushing overhead.

The blast hit me when I reached the gap between Hilbre and Middle Eye. Sand and grit was being blown at speed along the ground. At the southern tip of Hilbre, a large 4×4 with a high clearance was parked up. I waved at the lifeguard at the wheel, but was studiously ignored. A few minutes later, as I passed Middle Eye and embarked on the long stretch to Little Eye, I heard a distant engine behind me. The vehicle was coming my way, very, very slowly, flashing its hazard warning lights. I walked on. The 4×4 kept a strict 200 yards behind me. Was I being followed? Or just herded? I'd seen several other people still on Hilbre, so I wasn't the last to leave. Eventually the 4×4, giving me a wide berth, overtook me and purred off towards the mainland. I followed, enjoying the sun of the late afternoon, and the fresh breeze that had blown the morning's rain away. A few stragglers followed behind me.

When I returned once more to the Dee Lane Slipway, the lifeguards were shutting up shop. Anybody left on the island, or still crossing the sands as the waters rose, would have to fend for themselves. Later that evening, I looked out towards Hilbre, a black streak against a grey sky. The sea had returned, and Hilbre had become an island once more.

They say that Liverpool and the Mersey are detached from their hinterland: they do not look inward to the rest of England, but outward to America and the rest of the world. Something similar might be said of Hilbre. Facing Wales across the Dee, and Ireland across the sea, in its modest way Hilbre really does mark an edge of England. For the birds that blow through, though, Hilbre is only a brief staging post, not a border. Its territorial status is an irrelevance amidst thousands and thousands of miles of open sea, a whole world of restless, unbounded air.

All at Sea

The Farnes

An atmosphere of remoteness that clutches at your heart.

– J. H. Ingram, *The Islands of England* (1952)

The rain drummed on the roof all night. My heart sank. Had I made another long, wasted journey up the length of England? The previous year, planned trips to the Farne Islands had twice been frustrated by family illness, and then a third time by high winds, torrential rain and choppy seas – conditions in which the boatmen of Seahouses would not sail.

By morning the heavens had defied the forecast, and closed. The air was left thick with mist. The sailing I'd booked with Billy Shiel's Boat Trips didn't embark until 1 p.m., so I had the morning to explore the massive bulk of Bamburgh Castle, high on its rock, dominating this stretch of coast. Staring glumly from the ramparts through the greyness, I could just make out some flat shapes out at sea, shadows of what might have been Megstone and Inner Farne. Would the boat sail in such poor visibility? The shadows disappeared when I tried to focus on them through my camera – like those evasive subatomic particles whose momentum and position cannot be measured simultaneously. I took some comfort from the fact that the mist was low cloud, the remains of the night's deluge, rather than sea fret or haar, a curtain of cold fog sometimes drawn in from the North Sea along these coasts. Sea fret can squat over the land for days, depressing the temperature and the spirits. A few miles inland, the sun will be shining.

Were the Farnes going to elude me again? Perhaps they shared the evanescence of Eynhallow, a small Orkney island in the sound between Mainland and Rousay. Eynhallow is said to disappear from time to time, although this may have more to do with meteorological factors than supernatural agency. In Orkney folklore the island was the summer home of the Fin Folk, shape-shifting merpeople who on occasion took mortal spouses, a custom shared by the selkies, the seal folk, who at times took human form and human lovers.

Or perhaps the Farnes, today at least, were completely illusory, like those lands described and charted by early Arctic voyagers, but which turned out to be emanations of that most convincing of mirages, the fata morgana. In 1818 Commander John Ross of the Royal Navy, pushing far north up Baffin Bay in search of the Northwest Passage, turned back from Lancaster Sound, between Baffin Island and Devon Island, believing the way was blocked by a range of mountains, which he named the 'Croker Mountains' after the first secretary of the Admiralty. In other accounts, the passage blocked was Jones Sound, further north, between Devon Island and Ellesmere Island, and the range named the 'Barnard Mountains'. Such is the uncertainty of both history and geography at the edge of the known world – or at least the world known to European cartographers. Whatever the location and identity of Ross's mountains, they turned out to be no more than figments of atmospheric distortion, while Lancaster Sound proved to be the key to the Northwest Passage. Various other ephemeral lands west of Greenland and to the north of the Canadian mainland were reported throughout the nineteenth century, never to be seen again.

Despite their propensity to hide behind veils of mist and sea fret, the Farnes – an archipelago of small rocky outcrops off the Northumberland coast – have been documented for some 1,500 years, and have provided harsh homes for hermits, then light-house keepers, and now, from March to December each year, a handful of National Trust rangers, who monitor the birds and the seals. There are between fifteen to twenty-eight islands, depending on whether they are counted at high or low tide. The outcrops consist of dolerite, a hard volcanic rock known locally as whinstone, from the noise it makes when hit with a hammer. Some of the names of the islands embody the harshness of the geology: the Scarcars, the Harcars, Clove Car, Crumstone, Knivestone. The channels between them are known as 'guts': Wideopen Gut, Piper Gut, Middin Gut, Craford's Gut. There is a cleft in the northwest corner of Inner Farne called Churn Gut; during storms water is forced into this passage, and then jets through a small fissure eighty or ninety feet into the air. (I have seen a similar phenomenon on the west coast of Iona.) The islands were once linked to the mainland by softer limestone, which has since washed away; sea-level rises following the last ice age also contributed to their isolation.

The solidity of the Farnes has been left in no doubt by the many ships that have broken their backs on their rocks. The most famous of these was the *Forfarshire*, a state-of-the-art paddle steamer designed to carry both cargo and passengers. Built in 1834, it was commissioned by the Dundee and Hull Steam Packet Company to ply the east coast between the River Humber and the Firth of Tay. Previously, ships had found safe passage past the Farnes by sailing through the channel between Inner Farne and the mainland, an area free from submerged rocks. But newer ships such as the *Forfarshire* had too deep a draught for this shallow channel, and were obliged to negotiate around the Outer Farnes.

On 5 September 1838 the *Forfarshire* embarked from Hull for Dundee, with some forty passengers aboard, together with a crew of twenty-two. During the voyage, the ship developed problems with its boilers, and had to rely increasingly on its backup sails. By 11 p.m. on 6 September the ship had passed

Berwick-upon-Tweed, but then a gale-force wind came out of the north, driving the crippled ship back towards the Farnes. At 4 a.m. the following morning, the *Forfarshire* struck the rock of Big Harcar.

There were only three people that night in the lighthouse on Longstone, the outermost of the larger Farnes, beyond the Harcars and Clove Car. The three were the lighthouse keeper William Darling, his wife Thomasin and their twenty-two-year-old daughter Grace. Unable to sleep with the noise of the storm, Grace peered out of her bedroom window. She could just make out a dark shape on Big Harcar. It could only be a ship. She woke her father. By 7 a.m. it was light enough to see through a telescope that there were three or four survivors on the rock. The storm still raged. William and Grace decided they must mount a rescue. All they had at their disposal was a twenty-foot coble, an open boat powered by oars and sail. They launched it into the furious sea, and took a course south through Craford's Gut, and then westward in the lee of the Harcars. As they drew closer, they could make out that there were nine or ten people on Big Harcar. Drawing alongside, Grace kept the boat in place with her oars, while her father jumped onto the rock. William realised that there was not room in the coble for all the survivors. He was forced to separate a living mother from her two dead children, and to abandon the corpse of a drowned clergyman. The mother and two other passengers (one injured) were helped aboard the coble. They were joined by two of the *Forfarshire*'s crew, who helped man the oars on the voyage back to Longstone. Four men were left on the rock. Once the coble returned to Longstone, Grace disembarked with the bereaved mother and the two other passengers, while William and the two *Forfarshire* crewmen returned to Big Harcar to pick up the remaining survivors.

When the press got hold of the story, they focused on the role of Grace. She became an instant celebrity, while her father's role was largely ignored. Perhaps a virginal young heroine such as Grace was what the new Victorian age was looking for. Just the year before, the eighteen-year-old Victoria had succeeded to the throne; her coronation had taken place three months before Grace's rescue. Victoria brought with her a promise of innocence, in contrast to what many regarded as the disgraceful goings-on of her predecessors on the throne, her 'wicked uncles', George IV and William IV.

Visitors flocked to Longstone to catch sight of Grace. Among them were numerous artists, commissioned to capture her likeness and to depict the dramatic rescue. Prints of the results became best-sellers. The manager of the Adelphi Theatre in London invited Grace to play herself in his new drama, *The Wreck at Sea*. For £10 a night, she was to appear on stage for fifteen minutes, rowing a stage boat to a stage wreck, where she was to rescue the survivors. Grace declined the offer, but the play went ahead anyway. Five years later, William Wordsworth, newly appointed as poet laureate, produced a somewhat laborious poem:

> All night the storm had raged, nor ceased, nor paused,
> When as day broke, the Maid, through misty air,
> Espies far off a Wreck, amid the surf,
> Beating on one of those disastrous isles . . .

This effort was followed fifty years later by Algernon Charles Swinburne, by which time Grace was long dead, her place firmly secured in the English Valhalla:

> East and west and south acclaim her queen of England's
>   maids,
> Star more sweet than all their stars and flower than all their
>   flowers,
> Higher in heaven and earth than star that sets or flower that
>   fades.

William Topaz McGonagall also chipped in, with verses up to his usual standard:

And nine persons were rescued almost dead with the cold
By modest and lovely Grace Darling, that heroine bold;
The survivors were taken to the light-house, and remained
  there two days,
And every one of them was loud in Grace Darling's praise.

Grace Darling would not have recognised herself, later writing, 'I had little thought but to exert myself to the utmost, my spirit was worked up by the sight of such a dreadful affair that I can imagine I still see the sea flying over the vessel.' She was an ordinary and modest Northumberland lass, and turned down the many offers of marriage she received in the wake of her unlooked-for fame. Although born in Bamburgh, she had spent nearly all her life on the Farnes; at just a few weeks old, she was taken to Brownsman Island, where her father kept the lighthouse. Brownsman is one of the larger and more fertile of the Farnes. There were no trees, but the soil could support a small vegetable plot, and a few goats and sheep. Fresh water had to be brought in barrels from the mainland, as did dried pulses and grains. There were plenty of rabbits for the pot, and fish in the surrounding waters. The family made some extra pennies by selling eider ducks for the table, and their feathers for down to fill quilts ('eiderdowns'). Grace herself had a pet eider duck, and learnt needlework and housekeeping from her mother, and to read and write from her father. The Bible, poetry, history and geography provided approved texts; novels were frowned upon, while playing cards were 'the Devil's books'. William Darling also taught Grace and her brothers and sisters how to tend the light, and to sail and row the coble.

Brownsman may have provided a relatively comfortable home for the Darling family, but it was not ideally suited as the location of a lighthouse. To the northeast of Brownsman, the rocks of the Outer Farnes extend more than a mile further out into the North Sea. Trinity House decided that a new lighthouse was required on Longstone, the outermost of the larger islands. In 1826 the

Darlings moved to the newly constructed lighthouse (which is still in use, though now automated). Longstone was a barer, more rocky island, so the family still kept their vegetable plot and animals on Brownsman, rowing to and fro as required. On Brownsman, the Darlings had lived in a cottage beside the light; on Longstone, they cooked and ate in the ground floor of the lighthouse itself, and slept in the three bedrooms above; it was from one of these that Grace first saw the wreck of the *Forfarshire* in the dark of the storm.

Grace did not long survive her moment of glory. She died of tuberculosis in 1842, aged only twenty-six, tended towards the end by the Duchess of Northumberland (or her physician) in a house close to Alnwick Castle, the ducal seat. In Bamburgh churchyard there is not only the grave of the Darling family, but also an ornate catafalque for Grace herself. Inside the church there is a memorial in the form of a sarcophagus on which a full-length stone carving of the heroine reclines, her eyes closed in peace, an oar placed under one arm. 'Before you go,' a notice asks the visitor, 'remember to say a prayer for all sailors and those in danger on the sea.'

~

The light on Brownsman was not the first light on the Farnes. Permission was first given for a beacon in 1669, which was probably lit on Prior Castell's Tower, an early Tudor pele tower on Inner Farne, named after Thomas Castell, prior of Durham, who had responsibility for the monastery established on Inner Farne in the thirteenth century. Although no doubt the early residents adhered to a life of labour and prayer, in time the monks employed servants to till the soil and tend their livestock. Seals and porpoises were hunted, and eggs and eider down collected. By the time of Henry VIII, the rigours of island life proved too much for many of the monks, who preferred to spend their time on the mainland, 'haunting a woman's house over oft in the forenoon'. It seems that by the time of the Dissolution of the

Monasteries, the establishment on Inner Farne was already pretty much abandoned.

Before the foundation of the monastery, Inner Farne had been the resort of hermits. One such anchorite, St Bartholomew of Farne, lived there for forty-two years, until his death in 1193. He had been put out when he first arrived by the presence of another hermit on the island, a man called Aelwin. Bartholomew succeeded in getting rid of his rival by making himself as objectionable as possible; it was said he never washed. On the plus side, one modern hagiographer writes, his 'more endearing traits included continual cheerfulness, love of fishing, his fondness for his pet bird, and his generosity to his many visitors'.

The tradition of the Farnes as a home for hermits goes back at least to St Cuthbert, in whose cell, many centuries later, Bartholomew set up home. Cuthbert retired as prior of Lindisfarne in 676 and built himself a solitary cell on Inner Farne. It had just one window, through which only the sky could be seen, so Cuthbert would not be distracted from his meditations. Bede, in his *Life*, writes that before Cuthbert's arrival on Inner Farne, the only inhabitants were evil spirits, whom the saint succeeded in driving out. Bede also wrote that the stones of the walls of Cuthbert's cell were too heavy for a single man to lift, so concluded that the saint must have had the help of angels – who also brought him bread from heaven.

Cuthbert seems to have had a special relationship with wild animals. Bede recounts that while visiting the abbey at Coldingham near the Berwickshire coast the saint immersed himself in the sea up to his neck and spent the night in praising God. Only at dawn did Cuthbert emerge onto the shore, where he fell onto his knees to resume praying. At this point two otters 'came up from the sea, and, lying down before him on the sand, breathed upon his feet, and wiped them with their fur, after which, having received his blessing, they returned to their native element'.

On Inner Farne, Cuthbert grew his own barley, but was distressed that the gulls stole some of his grain. He put a stop to this by preaching them a sermon on the text 'Thou shalt not covet another's goods.' Filled with remorse, the gulls desisted. However, the saint was more than happy to share some of his grain with the eider ducks, who became his friends. Upset that the local people fed themselves on the eider and their eggs, he decreed that the birds should be protected (a prohibition that had lapsed by the time the Darlings took up residence). Ever since, eider have been known in the Northeast as 'St Cuthbert's birds', or, less formally, 'Cuddy ducks'.

~

When I arrived at Seahouses, the harbour was full of eider, a dozen or so mothers each with half a dozen chicks. The females are pale brown with numerous dark brown bars; there were no males, which have striking black-and-white markings. Perhaps they were floating in a group somewhere offshore in one of their distinctive 'rafts'. Hundreds of pairs of eider still nest on the Farnes.

Seahouses is a very different village from genteel Bamburgh, which will not let you forget that its castle was once the capital of the ancient Kingdom of Northumbria. With its neat stone cottages and pretty gardens, Bamburgh has an aura of green wellies and Land Rover Discoveries. In contrast, Seahouses, just down the coast, has fish and chips galore, amusement arcades, tattoo and piercing parlours, and the Farne Gift Shop, an Aladdin's cave of buckets and spades, model lighthouses, fluffy seal pups and toy puffins.

When I presented myself at Billy Shiel's kiosk at the harbour, I was told that I could not land on Inner Farne, the only island open to visitors apart from Staple Island. The National Trust, which now owns the Farnes, had decreed that the forecast of thunder and lightning made any landing too dangerous. However, I could take a cruise round the islands. The rain had long stopped, and there was little wind. Just the thinning mist remained.

So I joined a dozen or so other tourists aboard *Glad Tidings 6*, a twenty-foot open-backed motor launch. As we left the shelter of the harbour and the families of eider we hit the swell, which must have been three or four feet high. The waves seemed to come from two different directions, unpredictably, making us pitch then roll as the skipper calmly steered his course. It was not easy to stay upright – indeed, it proved impossible. Kneeling on the bench beneath the gunwale, I wedged myself against the rail, doing my best to take photographs. The persistent mistiness did not help.

At first there was just the odd gull, lazily flapping around. Then – one from behind, two from either side, one ahead – guillemots flew low and fast over the water, like fighter jets flying beneath the radar. Then there were more and more, some bobbing in flocks on the surface of the sea, some landing, some clumsily taking off with rapid wingbeats, frantically thrashing the water with their feet. With their short wings, sharp bills, white breasts, and black heads and backs, these large auks are striking birds.

A pair of gannets more slowly and purposefully skimmed the waves. If guillemots are the fighter jets of the seabird world, gannets are the long-range stealth bombers. These two might well have come all the way from the Bass Rock in the Firth of Forth, forty miles to the northwest. As the *Glad Tidings* motored on, other birds began to appear: razorbills, terns, puffins – masses of puffins, the comic tenors of the seas, with their clownishly striped beaks and their frenzied wingbeats, up to 400 per minute, enabling them to reach speeds of fifty-five miles per hour.

When we came to the low whinstone cliffs of the islands, they were white with guano, covering the black of the underlying rock. Kittiwakes nested on tiny ledges beneath the cliff tops, while all along the crests were rows and rows of guillemots, thousands of them packed shoulder to shoulder, their ranks every now and again punctuated by a shag spreading its wings. Guillemots do not make nests, instead laying their single egg on the bare rock. Their living conditions are so cramped that they have evolved various appeasement rituals with their neighbours, such as mutual preening.

Every square inch of rock was occupied by birds; both air and sea were full of them. When surveyed in 2012, the Farnes were home to 1,000 pairs of shags, 1,200 pairs of Arctic terns, 4,000 pairs of kittiwakes, 36,000 pairs of puffins and nearly 50,000 guillemots. Puffin numbers have since crashed, partly because warming seas have pushed the sand eels on which they feed further north, partly because of plastic pollution in the oceans. Many of these birds are great travellers. Kittiwakes and puffins spend most of the year far out at sea, guillemots may fly more than fifty miles from their breeding grounds each day in search of food, while Arctic terns undertake the world's longest annual migration: many will cover more than 50,000 miles each year, chasing the summer from the Arctic to the Antarctic and back. One Arctic tern ringed on the Farnes in 1982 reached Melbourne in Australia later that year, having flown 14,000 miles in three months.

The Farnes are also home to the only colony of grey seals on the east coast of Britain – 6,000 of them breed here, the pups being born in the autumn. As we motored around the islands, some of the seals were just heads in the water, peering at us with interest, while others were laying out on slabs of rock revealed by the receding tide, waving their flippers in a leisurely way, rolling over on their sides, basking in the few rays of sun that were just beginning to penetrate the mist.

In some ways I was glad I couldn't land. The Farnes belong to the seals, above all to the birds. They are all protected now. This wasn't the case in the past. In 1769 the Welsh naturalist and traveller Thomas Pennant visited the Farnes, on his way north to tour Scotland, and reported on the methods used to harvest seabirds on the Pinnacles of Staple Island: 'The fowlers pass from one to another of these columns by means of a narrow board, which they place from top to top, forming a narrow bridge, over such a horrid gap that the very sight of it strikes one with horror.'

The slaughter of young kittiwakes for their fashionable feathers began on the Farnes in 1802. Throughout the nineteenth century, the islands became a target for egg collectors and self-styled 'naturalists' with guns, in search of trophies to shoot and stuff. Some of the egg collectors were professionals, sending their loot to London to be sold. These predations eventually led to the formation of the Farne Island Association (FIA), set up to protect the birds in the breeding season. Up until 1922, one family owned the Inner Farnes, another family the Outer Farnes. In that year the FIA, with Lord Grey of Fallodon (former foreign secretary and a keen ornithologist), mounted a public appeal to raise funds to purchase the islands. They succeeded in 1925, and handed the Farnes to the National Trust.

We had an hour and a half cruising round the rocks and islets, stopping every now and again to come up close to the birds and seals, bobbing alongside the cliffs. Once we were back in Seahouses harbour, lightning split the sky, thunder rumbled and the heavens opened. The low cloud, having earlier thinned and risen, sank again and thickened more than ever, making for a slow and difficult onward drive up the A1 to Edinburgh. I took some comfort as I peered ahead into the murk from the thought of the tens of thousands of birds and seals secluded safely in their own world on the Farnes, noisily living and breeding and feeding behind a dense curtain of mist.

In Search of Lost Time

The Isle of Wight

She thinks of nothing but the Isle of Wight, and she calls it the Island as if there were no other island in the world.

– Jane Austen, *Mansfield Park* (1814)

. . . the Island is thoughtful and foreboding. Beautiful but isolated and a bit unhinged. Something is just *off*.

– Hannah Ewens, 'Home Coming: The Isle of Wight', *Vice* (28 September 2018)

The twenty or so miles that separate England and France across the Strait of Dover is not the same distance as the twenty or so miles from London to Dorking or Gravesend. Distance is relative. That between England and France is elongated by language and culture, by centuries of warfare. That between London and Dorking or Gravesend can be extended by traffic jams or by leaves on the line. But London and its surrounding towns are proximate in the mind's map, whatever time it takes to travel between them. They are not two countries divided by an unbridgeable moat of water and mistrust.

The distance between Portsmouth and Fishbourne on the Isle of Wight – England's largest island by far – is only seven miles. The ferry takes forty-five minutes to make the crossing. But some

The Isle of Wight

people say the Isle of Wight is separated from the mainland not by forty-five minutes, but by twenty years. Others give figures of thirty or forty. Locals maintain that the ferry across the Solent is, mile for mile, the most expensive sea passage in the world. The cost, they say, contributes to the Isle of Wight's isolation. An advertisement by one of the ferry companies reassures prospective visitors to the Isle of Wight that they won't need a passport. But you might need a diving bell to take you that far down into the past.

For me, the depth was sixty years, almost a whole lifetime. I'd last been to the Island in 1959. I wasn't sure I wanted to go back. My oldest sister, Patricia, said she wouldn't ever. In her mind it was a lost Eden: 'The "Isle of Wight" was always a place of perfection, locked away in the past, never to be re-visited for fear of a loss of innocence, of seeing tawdriness and suburbia instead of a lost paradise.' In her mind, the 'Isle of Wight', in its inverted commas, was not so much a place as a half-forgotten dream.

Every year when I was little, myself and my four older siblings would travel with my mother down from our home in Edinburgh to the Isle of Wight, to spend Easter with our maternal grandparents. In the 1930s my grandparents had built a holiday home on the Island (as they always referred to it, with a capital I), in St Lawrence on the south coast, in the area known as the Undercliff. They called the house Stragill, after their first marital home, a tiny cottage in Donegal. During the First World War my grandfather, a Royal Navy electrical engineer, had been posted to a submarine listening station on Lough Swilly. The young couple had little to live on, but my grandfather remembered his time there as blissfully happy. My grandparents had retired to the new Stragill in the early 1950s. My father would join us for the Easter weekend.

Compared to the sunless, skin-nipping chill of Edinburgh in April, Easter on the Isle of Wight was an interval of warmth, of greenery and flowers. I remember swathes of daffodils. Even by the sunny, southerly standards of the Island, the Undercliff is known for its warm microclimate. This attracted many wealthy

Victorians – particularly those with respiratory problems – to make their homes there. A vast TB sanatorium was built at Ventnor.

But the benefits of the climate came at a cost. The Undercliff is a narrow strip of relatively level land trapped between steeper slopes above and below, and is one of the most geologically un-stable parts of the British Isles. It was formed by a series of land-slides in the wake of the last ice age, and landslips are still not infrequent, as the heavy chalk and upper greensand cliffs of the Downs above press down on the softer clays and sands below. Roads can be wiped away, and many of the original Victorian villas have been destroyed by landslides, or have fallen into the sea.

Some Victorians preferred other, more stable parts of the Isle of Wight. The queen herself chose East Cowes, at the opposite end of the island from Ventnor, as the location of Osborne House, an extravagant palazzo in the Italian Renaissance style designed by Prince Albert. The couple had their own private beach on Osborne Bay, where Victoria kept her bathing machine (a luxurious affair, complete with plumbed-in water closet). Victoria continued to make visits to Osborne after Albert's death. There are accounts of her picking primroses in the grounds for her favourite prime minister, Benjamin Disraeli. (In 1817 the poet John Keats had opined that 'the Island ought to be called Primrose Island'.)

There had been earlier royal connections. In 1647, during the Civil War, Charles I was imprisoned in Carisbrooke Castle, outside Newport, the Island's capital. There is a story that he refused the offer of a file to saw through the bars of his window on the grounds that he was slim enough to slip between them. He wasn't, and within two years had been found guilty of high treason and beheaded.

Later royals were attracted to the Island for the yachting. In 1817 the Prince Regent (the future George IV) became a member of what was later known as the Royal Yacht Squadron, the organ-iser of the annual Cowes Week regatta. The Royal Yacht Squadron was at one time one of the snobbiest institutions in Britain. Even the future Edward VII, then Prince of Wales, was unable to persuade the committee to allow his friend, Sir Thomas Lipton, to become a member, until the last year of his life. The objection was that Lipton, although one of the most successful competitive yachtsmen of his age, was merely a 'grocer'. Women were only allowed to become full members in 2013.

Queen Victoria died at Osborne House in 1901. A few years earlier, in 1892, her poet laureate, Alfred, Lord Tennyson, had also died on the Island, in Faringford House near Freshwater, which had been his main residence since 1853. Like other in-comers, he delighted in the Island's air, which, he said, 'is worth sixpence a pint'. In a poem addressed to the Rev. F. D. Maurice, Tennyson wrote fondly of the Isle of Wight,

> Where, far from noise and smoke of town,
> I watch the twilight falling brown
> All round a careless-ordered garden
> Close to the ridge of the noble down.

The chalk ridge referred to, to the south of the house, extend-ing towards the Needles, is now known as Tennyson Down. At its highest point is the Tennyson Monument, a tall Celtic cross 'raised', according to the inscription, 'as a beacon to sailors'.

My grandparents, too, have a monument on the Isle of Wight, albeit a more modest one in the form of a gravestone in Sandown Cemetery. Commemorated too is my mother's older brother Tom, who in 1925 died on the Island of peritonitis following a burst appendix. He was just a few days short of his eighth birth-day. Perhaps my grandparents ended up living on the Island to be closer to their lost boy.

My grandmother – whom I barely remember – died in 1958, and my grandfather sold up and moved to Scotland in 1960, to be nearer his daughter. I was born in 1954, so my Easter visits to the Isle of Wight are at the very edges of my memory. I remember green buses, a terrifying ride on a donkey on a beach, the steep, sand-covered steps and metal handrail leading down from the road to the house, the flower-strewn lawns, the woods at the bottom of the garden, paths through lush vegetation down to rocky beaches and the sea. There was also the paradox of having to walk 'up' to the Downs. Then there was the thrill of hearing stories and seeing prints and photographs of dramatic landslides, collapsing cliff faces, ruined villas. I wonder whether this sense of imminent disaster penetrated through my daytime happiness into my dream life, because I remember that as a young child my sleep on the Isle of Wight was fraught with nightmares – of cold, unseen spectres brushing my cheek; of witches poking me with their fingers; of giant mandrills baring their teeth ready to bite my face. This last horror was inspired by a colour plate in my mother's childhood encyclopaedia of animals. The mandrill in the painting was in a fight to the death with a leopard. Even at that young age, I came to realise that Paradise had its dark side.

~

Most of this part of my past is lost, and what is left has probably altered unrecognisably in the process of rehearsal and repetition in my mind. The past is indeed another country, a hermit kingdom ringed by unbreachable borders. But my lost past, the lost past of sixty or so years ago, is shallow, not far beneath the surface of the present. At the foot of the Undercliff, as the accumulated layers of the island crumble and slide down to the sea, you might, if you're lucky, find traces of a much, much deeper past. The Island is one of the richest areas in Europe for fossilised dinosaur bones. Human memory – even the whole of human experience – is rendered insignificant. Our species has only been around for 200,000 years. Dinosaurs dominated the planet for 165 million years.

The pattern of land and sea on the earth was very different in the days of the dinosaurs. As the continents drifted round the planet, oceans formed and drained. This process continues to this day, but the movements are so slow that they are only noticed when two tectonic plates jerk against each other. Even a movement of a few inches can cause an earthquake.

Over a smaller timespan than it takes for continents to migrate, the most significant determinant of changes in coastline is the rise and fall of the sea, which is in turn dictated by climate. During the last ice age, when the seas were shallower, their waters locked up in the ice caps, the Isle of Wight was linked to the mainland in what is now Dorset by a ridge of chalk. The Solent – the channel that separates the Isle of Wight from the rest of England – was then not a channel of the sea, but a mighty river flowing east from Poole Harbour, eventually joining the even mightier river that then ran down the otherwise dry bed of what is now the English Channel.

But with the melting of the ice, vast volumes of meltwater flowed into the Solent. Sea levels rose, the umbilical ridge of chalk was breached, and the Solent became part of the sea. This occurred some time between 8,000 and 9,000 years ago. All that is left of the connecting ridge are the Needles off the western tip of the Isle of Wight, and the Old Harry Rocks off Handfast Point on the Isle of Purbeck – which is not an island at all, but a peninsula. Another factor was the process called post-glacial rebound (see page 165).

By the end of the last ice age, the Isle of Wight had more or less adopted its present shape, a large diamond set in the crutch of southern England. But this diamond shape is not hard and fixed like a diamond is. The rocks of the island – mostly clays, chalks, mudstones and sandstones – are slowly eroding, like a rotting tooth. Much of the south coast, particularly the long stretch known as the Back of the Wight, resembles the edge of some vast open-cast mine. It's as if the earth itself was in the process of being eviscerated by a fleet of giant but unseen mechanical diggers, with the red, grey and white guts of the Island being blasted out and scraped away. As the land continually collapses and spills down to the sea, fossils are constantly brought to the surface – fossils not only of dinosaurs, but also of gastropods, insects, crocodiles, turtles and mammals. Archaeologists have also found evidence of settlement by Mesolithic hunter-gatherers some five fathoms under the waters of the Solent, off the north coast at Bouldnor.

~

The island's name is not as mutable as its geology – but nor is it anything like as old. *Wight* is thought to be a Brittonic (Old Celtic) name, possibly meaning 'place of the division', a name that may reflect the island's geographical position, between the two arms of the Solent. When the first Celts arrived on the island and displaced or absorbed the earlier inhabitants is unknown, although Julius Caesar tells us that in 85 BC the island was settled by the Belgae, people belonging to a large confederation of tribes from northwest Gaul, who lived between the Rhine and the Seine. Their language was predominantly Celtic, though it may have included Germanic elements. The Romans, who conquered the Isle of Wight following their invasion of Britain in AD 43, called the island *Vectis* or *Insula Vecta*. They built no major settlements, although the remains of several Roman villas have been found.

After the Romans abandoned Britain, the Isle of Wight was subjected to another foreign invasion, this time of Jutes (from modern-day Denmark), who established the independent kingdom of *Wihtwara* (meaning 'men of Wiht'). According to Bede, two of the earliest kings of the island, in the early sixth century, were called Wihtgar and Stuf. The island remained stubbornly pagan long after the rest of Anglo-Saxon England had converted to Christianity. Wulfhere of Mercia conquered the island in 661 and forced the inhabitants to become Christian, but after he left they reverted to their old ways. In 686 King Caedwalla of Wessex completed his conquest of Wihtwara, and, according to Bede, 'endeavoured to destroy all the inhabitants' and replace them with his own (Christian) subjects. The Island became part of the kingdom of Wessex; and then, under Alfred the Great, was absorbed into England. In the twenty-first century, the Island again has a thriving pagan population, paganism today being the third most popular religion.

The fact that the Isle of Wight was an island did little to protect its inhabitants. Its position, close to mainland England, and on the front line facing continental Europe, made it particularly vulnerable to invasion. It was subject to intense Viking raids, and following the Norman Conquest it was ruled for 200 years by the de Redvers family, who treated the Isle of Wight more or less as their own independent fiefdom. As she lay on her deathbed, the last de Redvers ruler, Isabel de Forz, was persuaded to sell the Island to King Edward I, and the lordship of the Isle of Wight came under the control of the English Crown (Edward's son, Edward II, presented the lordship to his favourite, Piers Gaveston). The Island did not cease to be vulnerable to foreign invasion, however, and in the later fourteenth century was raided by fleets from Castile and France. Two centuries later, the French were to return in force.

On 18 July 1545, during the reign of Henry VIII, an enormous French fleet appeared in the Solent. Under the command of Claude d'Annebault, the Admiral of France, the fleet consisted of 150 'great round ships', 60 coasters and 25 galleys. Nothing like it had been seen in English waters since William the Conqueror's invasion fleet had appeared off the coast of Sussex. D'Annebault's fleet possessed massive firepower, and carried a landing force of 6,000 troops. The fleet anchored off St Helens on the northeast coast of the Isle of Wight. On 19 July the English fleet, headed by the *Great Harry* and the *Mary Rose*, sailed out of Portsmouth to do battle with the French. The encounter was inconclusive, best remembered for the sinking of the *Mary Rose* – not due to enemy fire, but to incompetence. The lowest row of gun ports had been left open after firing, so when the *Mary Rose* heeled over while turning about, water poured through the ports and the ship capsized, and, according to a contemporary, 'suddenly she sank'. There were only about 40 survivors among the 600 or 700 men on board.

D'Annebault now decided to mount an invasion of the Isle of Wight. This, he thought, would draw out the English fleet once more. He may have also wished to establish a base from which to invade the mainland. French troops made landings along a ten-mile stretch of the east coast, between Seaview (north of the French anchorage off St Helens) and Bonchurch (now part of Ventnor). In anticipation of this move, the island's militia had been reinforced by men from Hampshire and Wiltshire, and the garrison, now numbering perhaps 5,000, concentrated itself on two strong points: Bembridge Down, to the south of St Helens, and St Boniface Down, above Bonchurch. French attempts to take the heights were in the end thwarted, and D'Annebault ordered the evacuation of the island. By 28 July the French fleet had left the Solent. It was one of the last foreign invasions of English soil (but not quite the last, as in 1667 the Dutch briefly occupied the Isle of Sheppey).

Three centuries later, even after the defeat of Napoleon, anxieties about the intentions of France persisted. In the 1860s, in one of the most expensive (in real terms) projects ever undertaken by a British government, Prime Minister Lord Palmerston oversaw the construction of a series of coastal forts to protect the British coast against possible French invasion. A number of these sites are on the Isle of Wight, including Fort Victoria and the Needles Battery, while in the Solent stand the isolated round bastions of Spitsand Fort, Horse Sand Fort, No Man's Land Fort and St Helen's Fort. With the defeat of France in the Franco-Prussian War of 1870–1, the threat presented from across the Channel receded, and the forts became known as 'Palmerston's Follies'.

~

As we sailed out of Portsmouth's Gunwharf Quays into the murk and drizzle of the open Solent, we could barely make out Spitbank Fort. I was once more accompanied by my wife Sally and our friends Alice and Tom. Although it was July, there was no indication that we were heading for one of England's balmier destinations. Rain and low cloud accompanied us we landed at

Fishbourne and drove over the Island's hilly centre to Ventnor. But when we reached the south coast, the microclimate took over. The clouds parted, and the sun came out.

We were staying in what had been the Seacliff Private Hotel and Boarding Establishment, 'Facing due south,' said an old advertisement on the wall, 'centrally in best part of Ventor': 'Owing to its elevated position, 150 feet above the bay, uninterrupted balcony and bedroom views are obtainable of the ship-dotted English Channel; the complete panorama from East to West embracing the Pier, Bathing Beach and beautiful Undercliff Coast Line. Special Winter Terms.'

The pier is long gone. The coast is disintegrating. And the three-storey hotel is now divided into apartments. Our Airbnb occupied the first floor, with French doors opening out onto a balcony with a wrought-iron balustrade, looking down, as promised, onto the beach below. There were old photographs of men in boaters and striped blazers accompanying women with big hats and parasols on the pier that is no longer there.

Below the balcony, across the narrow street, there was a group of young teens half-sitting, half-crouching on the wall of a belvedere overlooking the beach. The girls wore black skinny jeans, grey tops and long, straight hair. One had a brand new handbag nearly half her size on the wall next to her. Two of the boys had their hoodies up, the third had an Emo look, a long bang of dyed black hair hanging limply over his eyes. The girls sat side by side, while the boys restlessly changed position. One of the boys never took his hood down. School was over for the day, but not yet out for the summer. This was the place to be, on a terrace above the sea, between the strictures of the classroom and the confines of home. There was nothing especially different about these Island kids, apart from their ethnic homogeneity: the Island is 97.3 per cent white.

'British, white, poor.' That's how, in 2016, the then chair of Ofsted, David Hoare, characterised the population of the Isle of Wight. He went on to decry it as one of the worst performing local authorities, in educational terms, in the country. 'It's a ghetto; there has been inbreeding,' he told the *Times Educational Supplement*. He went on digging: 'There is a mass of crime, drug problems, huge unemployment.' The Island's MP described the comments as 'inaccurate, insulting and extremely unhelpful'. Hoare, a former City banker, subsequently apologised for his remarks, and resigned.

Hoare's observations didn't chime in with my own. Upstairs on a bus the next day we sat near some young teens on their way home from school. They were a lively, bright lot, sparking off each other and the world. I would have been proud of them if they'd been my kids.

There is an Islander identity. On the Island the natives call incomers and mainlanders 'overners' (i.e. 'overlanders'), a sub-species of whom are 'DFLs' ('Down From London'). In their turn, the inhabitants are known on the mainland (or 'North Island' as the locals sometimes jokingly call the rest of England) as 'caulk-heads', referring to the Island's past industry of caulking, the process of making a wooden boat waterproof by driving fibrous material between the planks. The intended insult is clear, but the Islanders seem to take the name as a badge of pride, responding that at least if they're thrown into the sea, they will float. In honour of the natives, a genus of pterosaur (an order of flying reptiles from the time of the dinosaurs), whose type species was found on the Isle of Wight, has been named *Caulkicephalus* (Greek *kephale*, 'head').

The Islanders do not appear to be insular. After all, the biggest industry is tourism. Although the Isle of Wight voted 62 per cent to leave the EU, I saw more European flags flying than Crosses of St George. Outside the theatre in Ventnor hung a rainbow banner. The Island has had its own flag since 2009: it features a diamond-shaped map of the of the Isle of Wight, floating above some stylised waves.

Ventnor is a pretty, genteel sort of place, the sort of place you find Hurst the Ironmonger, Lady Scarlett's Tea Parlour, a firm of solicitors called A. J. Careless and, on the promenade, Blakes (est. 1830), whose business is to hire out deckchairs, windbreaks and umbrellas. Along Ventnor's roads, contouring round the steep slopes of the Downs, there are elegant, stuccoed terraces, with bay windows and balconies. You half expect to meet, round every corner, some blazered and boatered young beau escorting his sweetheart to a concert on the promenade. But then you real-ise you are grasping at ghosts.

Ventnor was visited by a range of eminent Victorians and Edwardians, including Dickens, Elgar, Macaulay and Dr Thomas Arnold. Karl Marx convalesced here in the winters of 1882 and 1883. It did him no good: he died within weeks of his last visit. The Russian novelist Turgenev stayed for a while, but was thrown out by his landlady for excessive smoking. Both Mahatma Gandhi and the Emperor Haile Selassie holidayed in Ventnor, the former as a break from his law studies in 1890, the latter in 1938, while exiled from Abyssinia.

Curious to see where the French and English had last fought on English soil, on St Boniface Down above Ventnor, I walked with Tom up the town's steep streets past stone cottages with pretty cottage gardens. As we panted upward, I wondered why so many older people had chosen to retire to the Isle of Wight: 24 per cent of the population is aged sixty-five or over, compared to 16 per cent across the whole of Southeast England. You have to be fit to live here. Let alone fight here, as Sir John Fyssher, commander of the Hampshire militia, found to his cost on 21 July 1545. According to John Oglander, the seventeenth-century antiquary, when the English were forced by the French to retire, Fyssher, 'a fat gentleman ... not being able to make his retreat up the hill ... cried out £100 for a horse, but in that confusion no horse could be had, not for a kingdom'. The English were, it seems, routed, and 'not any family of note in the island', Oglander writes, 'but lost a father, brother or uncle'. The English rallied on top of St Boniface Down, and the next day when the French attempted to take their position they were repulsed. One French commander, the Chevalier d'Aux, was wounded in the knee with an arrow. His men fled, 'whereupon', Oglander tells us, 'some country fellow clove his head with a brown bill'. The French had had enough. They withdrew from the Island, and headed for home.

St Boniface Down is, at 791 feet, the highest point on the Isle of Wight. It is named after an early missionary to the island, originally known as Winfrith, who is said to have preached from Pulpit Rock on the flank of the Down in 710. On its seaward side the Down is thickly clad with holm oak, scrabbling up towards the top. Somewhere in the impenetrable thickets is St Boniface's (or St Bonny's) Well, now dry, a place people formerly visited to make wishes. Our path took us up a steep ridge of short-cropped grass alongside the woodland, past a number of feral goats. At the top, both woodland and grass give way to heath, with bracken, gorse and heather. Inside a high perimeter fence stand two tall masts, marked on the map as 'radio station' and 'radar station'. The modern radar station is involved in marine navigation; it is the successor to the RAF radar station built here just before the Second World War. Radar then being top secret, the purpose of the three tall pylons, each 365 feet high, remained a mystery to the civilian population, but the Germans must have had a suspicion, because in 1940 the pylons were destroyed by Stuka dive bombers. Before that, they had played a key role in the Battle of Britain. They were subsequently repaired, and the station was operational again by May 1941, and remained in use up to the 1960s. One of the buildings, the old receiver station, has recently been converted into a strikingly modernist holiday home, with plate glass windows looking far out to sea. It looks like the hideaway of some Bond villain. Beneath it all, there is a 1950s nuclear bunker, with walls ten feet thick and a ceiling designed to withstand 2,000 lb bombs. Three millennia ago the Down was home to many Bronze Age burials. Few traces of the ancient barrows remain, but you can see the bases of the radar masts, an abandoned pillbox and a scattering of other brick-built ruins around the perimeter – all now listed by English Heritage.

~

That evening we were to eat on the seafront, at the Spyglass Inn on the west side of Ventnor Bay. The pub flies the skull and crossbones, and customers are greeted by a larger-than-life, gaudily painted statue of a one-legged pirate. He has a parrot on his shoulder. As we walked along the esplanade above the empty beach, there was a sudden disturbance below us. From under the sea wall, a lone figure whooped and giggled, picking her way across the shingle towards the sea. It was a plump, curvaceous young woman in a swimsuit. She was carrying a very large, brightly coloured rubber ring. She might, if differently attired, have featured in a painting by Rubens, but her true home was undoubtedly a saucy seaside postcard by Donald McGill. (In 1953 a number of shops in Ryde, further up the east coast of the Isle of Wight, had been raided by police for stocking his work. Ryde now has a museum dedicated to the artist.)

On the edge of the shore, the young woman paused, one hand clutching the rubber ring, the other held elegantly out to the side, the fingers pointed. She stood before the tiny waves for a while, a Venus contemplating the sea from whence she'd come. Then she plunged in, splashing and whooping. 'There's a lovely sandy bottom here,' she didn't say (but one of McGill's buxom bathing belles had). From somewhere behind us there were cheers. 'She'll have done that for a bet,' said Tom. 'I wonder how many vodkas it took.'

Hannah Ewens, a prize-winning young journalist from the Isle of Wight, has written about her teenage years on the Island. At that age, she says, as her perspective widened, everywhere on the Island became a potential party spot. She remembers going to beach after beach, smoking weed, fights with urine-filled supersoakers, awkward fumblings. Then she remembers, perhaps like

McGill's large woman on the postcard, 'getting driven home with sand in your pants'.

~

Alice wanted to sit on the Wishing Seat again. She'd sat there more than a quarter of a century before. On that occasion she'd wished for a baby. It had worked. So she had a sentimental attachment to this lump of rock, deep in the woods of the Landslip, the area of the Undercliff between Bonchurch and Luccombe Bay. The Landslip has been an unstable area for thousands of years. The year 1810 witnessed a particularly violent landslide. Thomas Webster left the following account:

> I was surprised at the scene of devastation, which seemed to have been occasioned by some convulsion of nature. A considerable portion of the cliff had fallen down, strewing the whole of the ground between it and the sea with its ruins; huge masses of solid rock started up amidst heaps of smaller fragments, whilst immense quantities of loose marl, mixed with stones, and even the soil above with the wheat still growing on it, filled up the spaces between, and formed hills of rubbish which are scarcely accessible. Nothing had resisted the force of the falling rocks. Trees were levelled with the ground; and many lay half buried in the ruins. The streams were choked up, and pools of water were formed in many places. Whatever road or path formerly existed through this place had been effaced; and with some difficulty I passed over this avalanche which extended many hundred yards.[*]

Today the upper slopes are relatively stable, and support a density of trees and undergrowth that is almost tropical. The

---

[*] Quoted in Sir Henry Englefield, *A Description of the principal picturesque beauties, antiquities and geological phenomena of the Isle of Wight* (1816).

Victorians turned the area into what they called a 'pleasure ground', opening up a number of scenic paths, and promoting the Wishing Seat, plus other features such as the Devil's Chimney and the Chink. It all helped the Island's burgeoning tourist business.

The day being hot, we were glad of the shade of the canopy. The trees – mostly oak, ash and beech – were draped with ivy, mistletoe and old man's beard, while the ground was thick with bracken, brambles, bindweed and hart's tongue fern. Every now and again we'd get a glimpse of the sea, today a startling Aegean blue. The slopes leading down to the shore were still clearly active, the bare, hard clay stuck with dead sticks of trees. There were warning signs here and there, advising that a path had been closed 'due to erosion and unsafe conditions'.

We found the Wishing Seat. It was difficult to miss: at the side of the path a sign saying 'Wishing Seat' was placed at the foot of a gnarly mass of mossy, lichenous rock, vaguely in the shape of a chair. There was a mysterious cleft running from side to side through what would have been the back of the chair if it had been a chair. We took it in turns to sit on the seat, each making a private wish before continuing on our way.

And so, past Luccombe Chine, Knock Cliff and Yellow Ledge, we came to the town of Shanklin.

Shanklin has changed since Keats stayed here with his friend James Rice in July 1819, exactly 200 years before my own visit. At the beginning of his stay he wrote to his muse, Fanny Brawne:

> I am now at a very pleasant cottage window, looking into a beautiful hilly country, with a glimpse of the sea; the morning is very fine. I do not know how elastic my spirit might be, what pleasure I might have in living here and breathing and wandering as free as a stag about this beautiful coast if the remembrance of you did not weigh so upon me

Shanklin is now a bucket-and-spade, kiss-me-quick kind of place, very different from genteel Ventnor round the corner. It must have been here that I, aged three, had my first and last ride on a donkey. All I can remember is my terror; it was a long way down to the sand. There was once a pier with a theatre at the end, where Arthur Askey used to perform, but the pier was destroyed by the Great Storm of 1987. Today, if you want a pier, you have to walk further up the coast, to Sandown.

The beach at Shanklin was covered with windbreaks and sunshades, deckchairs and sun loungers. On one of the loungers an old man in shorts, dark glasses and bucket hat snored, open-mouthed. Two little boys were doing their best to save their sandcastle from the incoming tide. You could hire pedalos and paddle boards, go for a banana ride or take a jet ski tour. Along the esplanade there were seated shelters, roofed in the Anglo-Chinese style, and Jungle Jim's Summer Arcade. There was the Pirate's Cove Fun Park, with a huge model shark hanging by its tail from a sort of gallows at the entrance. There was a crazy golf course inhabited by several half-size dinosaurs. A few mums and dads slunk around despondently, wondering where their balls had gone. No one seemed to be playing with any enthusiasm.

The main part of Shanklin is at the top of a cliff, which separates it from the beach and the esplanade. This cliff-top Shanklin is a world of hotels, B&Bs and guest houses, still quiet before the school holidays. There were few people around, even on the main streets, which are dominated by burger bars, bookmakers and charity shops. Away from the beach, there is an impoverished, moribund air to Shanklin, similar to that found in many other English seaside towns since the Costas took their trade. The streets were so quiet I had to go into a Boots to ask where we could catch a bus back to Ventnor.

Two hundred years earlier, by the end of his month in Shanklin, Keats was similarly downcast:

The Isle of Wight is but so so &c. Rice and I passed rather a dull time of it. I hope he will not repent coming with me. He was unwell and I was not in very good health: and I am afraid we made each other worse by acting on each other's spirits. We would grow as melancholy as need be.[*]

The following February he began to cough up blood. 'That drop of blood is my death warrant,' he said to a friend. He was dead within a year.

We were glad to be back in Ventnor. From Shanklin the view is back towards the mainland. There doesn't seem to be any possibility here of escape from the dreary, everyday world of Little England. You're hemmed in, stuck in a past that has gone, and has no future.

From our balcony in Ventnor that evening we looked out to the southwest over the limitless sea, watched distant ships approach and cross the horizon towards France and the open Atlantic. As it grew darker, the atmosphere shimmered and blurred, opening possibilities, blending sea and sky in hazy bands of blue and grey and golden brown.

~

I could not leave the Island without seeing Stragill again. Could I bridge those sixty years?

I knew Stragill didn't have the same name any more. Twenty years before, my sister Alison had visited the Island, determined to find it. She'd narrowed it down to two neighbouring houses in St Lawrence's Undercliff Drive. At first she thought the house called 'High Trees' was the one, but further research convinced her that 'The Spinney' next door was the place we'd spent our childhood Easters. Her conclusion was the result of a coincidence of the sort my grandfather was always encountering (he put it down to Providence, I suspect). Someone had suggested she visit the Ventnor Heritage Museum, to see if they had any records. While she was there, she was overheard asking about Stragill by a woman who'd just popped into the museum. It turned out that

---

[*]   Letter to C. W. Dilke, 31 July 1819

this woman, Fay Brown, had been evacuated at the age of ten from Ventnor to St Lawrence, where her father rented Stragill in 1940–1. He had decided to move his family after the Germans bombed the radar pylons on St Boniface Down. At this time, my grandparents were living in Liverpool, through the worst of the Liverpool Blitz. Fay told my sister how exciting it was for her as a child to watch the dogfights above my grandparents' holiday haven, as the skies were scraped with arcs of vapour, smoke and flame.

You can walk along the coast between Ventnor and St Lawrence, sometimes along the shore, sometimes along the verdant terraces of the Undercliff. In places you are sent on a diversion because the old path has collapsed. St Lawrence itself, with cliffs above and cliffs below, is today a scattering of bungalows and mansions (some derelict) hiding among the trees. It doesn't feel like a co-herent village – although it is a much older settlement than Ventnor. The medieval church, dating from the twelfth century, is said to be the smallest parish church in England.

Walking west along Undercliff Drive, Sally and I were surprised how quiet it was. After all, this is the A3055, the main road along the south coast of the Island. We passed a number of bus stops. All said they were no longer in use. Then we realised that the main road ahead had been cut by a massive landslide some years before. You can barely see the break in the road on the OS map, but on the ground there is a mighty gulf. There didn't seem to be any intention to restore the road. The authorities must have concluded that it would be pointless: the Island here is slowly but surely slipping into the sea. So the Spinney today is in a much more peaceful location than Stragill had been when my grand-parents lived there. Undercliff Drive is now a dead end.

I opened the picket gate and followed Sally down the steep, narrow steps. There was a wall on the uphill side, and a metal rail down the other. It was as I remembered it, except now there was no sand on the steps. I don't know why there would have been sixty years ago. There probably wasn't. My mind had perhaps confused two separate memories. Unless one of us children had brought back a bucket of sand from the beach and spilt it here. If so, it had all been swept away.

On the uphill side, the house was a bungalow. On the downhill side, as I remembered it, there were two storeys. There was a man painting the front door, dressed in overalls. Perhaps the owners were out. I said hello, explained why we were there. 'Oh, I live here,' said the man, who turned out to be a lad. 'I'll get my dad. I've just finished my GCSEs, so I get to paint the house.' A tall man came to the door, with a friendly dog. I explained myself again. 'Oh, do come in, have a look round, excuse the mess. Sorry, I'm just on a work call. I've got to take it. But feel free.'

The inside of the house was no doubt very different to how it had been in the 1950s. There was now a large open-plan kitchen-diner. I half remembered lots of heavy wooden furniture, glass-fronted cabinets, dark bookcases, although I know I was remembering the furniture that accompanied my grandfather up to Scotland. What hadn't changed was the amount of light flooding through the windows from the south.

We walked down more steps at the side of the house. There were raised beds full of vegetables. My grandfather, a keen vegetable grower, would have been pleased. There was also a small summer house, with a somewhat abandoned look. We'd called this summer house the Chalet when we were young. It was filled with half-broken toys from my mother's childhood in the 1920s. My third sister, Patricia's twin Pamela, remembers it 'as our own special den', although only my brother Richard, the oldest of the five children, was allowed to sleep there. It was his very own grown-up adventure.

Fay Brown had given my sister copies of some photographs of the exterior of Stragill when she had lived there during the war. There have been many changes. The front of the house is now mostly panelled with irregularly lapped boards of bare wood. In Fay's photos, the only such boarding was beneath a first-floor bay window that no longer exists.

The lawn sloping beneath the house has shrunk, not just because I have grown. The garden used to extend further down the hill, to what we called 'the spinney', but half of it was sold off in the 1960s for further building. For Pamela, the garden was 'the most perfect place imaginable'. Patricia remembers a steep grassy bank covered in celandines, primroses and garden primulas, 'inspiring me to concoct a murky dye from their petals to colour a white handkerchief a slightly disappointing off-grey'. Beyond the spinney there was a little path that wound its way through the 'ivy-choked, magical forest of the Undercliff' down to the rocky shore of Binnel Bay. Patricia claims to have little grasp of the actual geography of the Island. 'I'll stay in my fuzzy Eden,' she says.

Returning to Ventnor along the top of the Undercliff, we came across a wooden sign. We were, it said, following footpath V71, otherwise known as Paradise Walk.

~

I hadn't quite reached the bottom of Memory Lane. For me aged three or four, the most magical place on the Isle of Wight was Blackgang Chine. Opened in 1843, Blackgang Chine is Britain's oldest amusement park. 'Black gang' is a 'black path', referring to the path that once descended the spectacular gorge here; a chine, on the Isle of Wight, is a steep-sided ravine cutting down through cliffs to the shore. Local folklore has it that long before the establishment of the amusement park, a giant lived in a cave in Blackgang Chine, which was then a verdant valley, covered with flowers. This idyllic spot was only spoilt by the giant's habit of feeding on the flesh of children. The giant's cannibalistic career came to an end when he and his home were cursed by a holy man with the following words, as recorded by Abraham Elder in *Tales and Legends of the Isle of Wight, with the adventures of the author in search of them* (1839):

> Nor flowers nor fruit this earth shall bear,
> But all shall be dark, and waste, and bare!

Nor shall the ground give footing dry
To beasts that walk, or birds that fly;
But a poisonous stream shall run to the sea,
Bitter to taste, and bloody to see!
And the earth shall crumble and crumble away,
And crumble on till Judgement Day.

The 'poisonous stream' referred to was the outpouring of a chalybeate (iron salt) spring, which in the 1830s had begun to attract visitors to the area, anxious to take the waters for their health. In 1839 a hotel was built at Blackgang Chine, and shortly afterwards a man called Alexander Dabell recognised the potential of the site for tourism. Dabell had moved with his family to the Isle of Wight (which he referred to as 'this far flung barbarous climb') from Nottingham, and had tried various ways to turn a penny, from selling hair oil to opening gift shops. Hoping to extend his businesses, he took out a lease on Blackgang Chine, defining the extent of his new property as the distance he could throw a stone. He set about building pathways down the ravine and laying out pleasure gardens. When a large fin whale was stranded off the Needles, Dabell bought it at auction, had the carcase flensed and the bones bleached, and displayed the skeleton at Blackgang Chine. It can still be seen there.

Dabell's original pathways and gardens have, like much of the coastline along the Back of the Wight, collapsed into the sea, and the present owners (Dabell's descendants) are continually having to move their attractions further inland. I half remember various tableaux involving large gnomes; if you placed a penny on a gnome's proffered hand, he would raise it to his mouth and swallow it. My sisters remember some 'crazy distorting mirrors'. All these memories were swept away by landslips long ago, although there is still, at the top of the cliffs, a hall of mirrors. This is now fenced off, and there's a sign saying 'Warning risk area. No unauthorised access.' Today, surviving attractions include a pirate play ship, animatronic dinosaurs, a haunted house, a cowboy town and a fairy village. To enter this 'Land of Imagination', you have to walk between the legs of a giant smuggler with a barrel on his shoulder, and pay £25 per head at peak times, unless you're under the age of four, in which case you get in free. We decided to pass.

Just west of Blackgang Chine we turned off down a side road leading towards the sea. The road came to an abrupt end. A wall had been built across it, and a large red sign said ROAD CLOSED. The road wasn't so much closed, as no longer there. Above the barrier stood the ruin of a two-storey house, its windows broken, its façade cracked. Wild honeysuckle was spreading over what had once been a terrace overlooking the sea. Concrete slabs from the terrace littered the grassy slopes below, joined by a tumble dryer and an old fridge. A few yards down the slope the grass ended at an ill-defined cliff edge, where the land turned to a kind of slurry-like grey clay, neither solid nor liquid. Flows of this indeterminate material had washed in waves and runnels down to the sea. It looked like the land was melting. For now, the flows were motionless, but the next heavy downpour would set them slipping again. To the east, the top edge of the fast-eroding clay was eating back towards the amusement park. It looked like the next attraction to go would be the rollercoaster called, appropriately enough, Cliffhanger.

~

There were two other places on the Island I wanted to revisit: Alum Bay, and the Needles. In my recollection, Alum Bay was a wide beach backed by huge cliffs streaked with bands of multi-coloured sand, towering impossibly and gaudily above me in the sunshine. My older brother had had a precious souvenir of the place, a test tube filled with layers of sand of different colours: pink, red, yellow, orange, white. I was allowed to look, but not to touch. Today, you are no longer permitted to take any minerals from what is now a Site of Special Scientific Interest.

We parked above the wrinkled, decaying cliffs of Alum Bay. Close to the car park was a garden filled with toy creatures – giraffes, toads, a Dalek, Tigger, a dinosaur, Noddy's friend PC Plod. But the majority of the residents were gnomes. Most of them were old and careworn. One shook his fist. They might all have been exiles from Blackgang Chine, figments of the imaginations – and perhaps nightmares – of generations of children.

As we walked along the Isle of Wight's southwestern peninsula towards the Needles, we could see across the glittering water of Poole Bay a distant dark streak: the Isle of Purbeck on the mainland of Dorset. All along the foot of the land there was a band of pale haze. It was as if the Isle was floating low in the sky. It could have been its own shadow.

I'd seen so many photos of the Needles – a line of three fin-shaped chalk stacks extending out to sea – that it was no longer possible to say whether I held any faithful images of what I'd seen as a small child. I'm not even sure if, back in the 1950s, you could get close to the Needles, as the whole of the end of the peninsula was then owned by the Ministry of Defence. It is now in the care of the National Trust. The Old Battery, built in the 1860s at the top of the 250-foot chalk cliffs at the tip of the peninsula, looks down on the Needles. The New Battery, further up the hill, was built in the 1890s, after concerns were expressed that the rock on which the Old Battery was built would collapse under the shock of firing heavy guns. The purpose of the batteries was to defend the westward approaches to the Solent and the naval dockyard at Portsmouth, but the only action either battery ever saw was firing anti-aircraft guns in the Second World War. After the war, both batteries were mothballed.

But with the warming of the Cold War in the 1950s, the New Battery was selected as a rocket-development site. The first rocket to be tested here was Black Knight, a guided missile. This was followed by Black Arrow, which, on 28 October 1971, became the first and only British rocket to launch a British satellite into orbit. This satellite, called Prospero, was operational for a couple of years. Among its roles was to measure the frequency of micro-meteoroids heading earthward. Prospero continued to be contacted annually for another quarter of a century. Signals from it could still be heard many years after it was officially deactivated, echoing the final speech of the original Prospero, the sorcerer-lord of the unnamed isle in Shakespeare's *Tempest*. At the end of the play Prospero gives up his magical powers, and pleads with the audience to be set free from his 'bare island':

> Now my charms are all o'erthrown,
> And what strength I have's mine own,
> Which is most faint . . .

Prospero the satellite is still up there, a tiny artificial island orbiting our small speck in space, taking 103.3 minutes to make one revolution of the earth.

~

It may have taken me sixty years to return to the Isle of Wight, but it was difficult to leave. There is something about the Island that's at right-angles to the rest of the world. Perhaps it's the time lag behind the mainland, perhaps it's the timelessness, perhaps it's the conflation of many separate times into a single present – as when a figure from a naughty 1950s seaside postcard appears before you on the beach, shrieking with laughter; or when you hear above you what sounds like a Spitfire, and then you look up and it *is* a Spitfire, looping the loop, and it turns out that this is a specially adapted two-seater Spitfire that takes tourists for very pricy thirty-minute spins over the Back of the Wight; or when the soft south coast of the Island sheds another layer of sand and clay like a snake sloughing its skin, and the remains of long-extinct reptiles tumble into the here and now.

The Island has a rough magic, an inexplicable magnetism. It's said that people brought up there find it hard to stay away, however much they've yearned to escape. Hannah Ewens has written about how she and her friends 'always end up looking past the sea, feeling lost or furious' at everything that keeps them there. Those teenagers who most waved their fists at the waves and who left with much fanfare when they reached the age of eighteen, she says, 'become shellshocked and return quietly. The Island, it makes us weird, you see.'

I felt flat on the train back to London. I'd spent a few days travelling back in time, reversing the passage of decades, and now I had to resume the journey in the direction of my own mortality. The Island itself was also on a one-way journey, a victim of entropy, heading towards disorder and dissolution as it washed away into the sea, dreaming itself into oblivion.

No Man Is an Island

An Afterword

No man is an Island, entire of itself; every man is a piece of
the Continent, a part of the main; if a clod be washed away by
the sea, Europe is the less . . .

> – John Donne, 'Meditation XVII',
> *Devotions upon Emergent Occasions* (1624)

Geography, like nature, ignores man-made borders. Migrating
birds acknowledge the authority of no sovereign state, nor does
water, nor do soil or silt or stones. The boundaries of maritime
islands are fluid, shifting with the tide. Although islands
properly have an area permanently above the high water mark,
there are many stretches of land that only appear twice a day –
gravel drifts such as the scars or skears of Morecambe Bay;
mudflats like Maplin Sands by Foulness, or the oozes of the
Medway; submarine banks such as the Goodwin Sands off the
east coast of Kent, the graveyard of innumerable ships, where
every summer the English defy the tide like King Canute by
organising a game of cricket at low water before the sea stops
play and King Neptune reclaims the sands once more for his
own. No nation-state has such fluidity built into its self-
conception; but the control over their own borders that
maritime states demand is undermined by coasts that that are
incessantly in motion.

There are processes much slower than the tides that alter the coastline, not only of England's islands, but also of the mainland itself. The end of the last ice age, which saw the inundation of Doggerland and so broke Britain's land link to mainland Europe, also had consequences for England's shores. The release of un-imaginable masses of water from the melting ice sheets caused sea levels to rise significantly around the world (a process that has accelerated with man-made climate change). But in Britain, another process was also at play – the process known as post-glacial rebound (see p. 165), a process that has seen swathes of southern England plunged once more under the sea.

There are haunting reminders of this loss still to be found along the south coast. When my children were young we would sometimes visit friends in Hastings, and, if it was summer and the day sunny, we would seek out the nearest bit of beach. Beyond the top of the crumbling sandstone cliffs that extend eastward from Hastings, the land once more comes down to the sea, at a place called Pett Level. At high tide, below the sea wall, the shore is a slope of shingle. But as the tide recedes the uncovered ground flattens, and we would walk out with buckets and spades across wave-rippled flats of mud and sand – some sand, but mostly mud. It was disappointing as far as beaches go, more a mire than a strand. The warm mud, grey and sticky, squeezed up between our bare toes. The barnacles and mussels clinging to small lumps of rock were slimy with the stuff, even the crabs seemed sluggish. There was an aura of decay, of life extinguished. Here and there we came across bits of wood, but it too was soft, unhealthily spongy, like rotting flesh. You could poke your finger into it. Then, further out, we found more and more remains – roots, fallen branches, tree stumps. Had this mudflat, this lost, forgot-ten place, once been forest?

It turned out that it had. Until around 6,000 years ago these tidal flats had been an open woodland, the oak and birch and hazel interspersed with wetland plants: sedges, marsh marigold, watercress, spearwort, bulrush, cowbane. Then the sea had been a hundred feet lower than it is today. In a small cave in the cliffs overlooking Pett Level, archaeologists have found fine flint blades, suggesting that Mesolithic hunters once looked out for game grazing between the trees on the plain below.

Those sandstone cliffs between Hastings and Pett Level are far from static. Every now and again a great block, undercut by wave action below, falls down into the sea. Along much of the eastern and southern coasts of England, the rocks are similarly soft and unstable, easily crumbled and worn away by sea and weather. In many places cliff-top dwellings have inched closer and closer to the edge, until eventually succumbing to gravity. Even more dramatically, whole towns have disappeared. In the Middle Ages, Dunwich in Suffolk was a port of a size to rival London. Then, in the thirteenth and fourteenth centuries, it was hit by a succession of storms and storm surges, destroying its harbour, most of its houses and all eight churches. Local legend has it that at certain tides the sound of church bells can be heard coming from beneath the waves. 'Dunwich, with its towers and many thousand souls,' wrote W. G. Sebald in *The Rings of Saturn*, 'has dissolved into water, sand, and thin air.'

Some have suggested that Dunwich's annihilation may not have been quite so dramatic, but rather the result of more gradual coastal processes that shifted the mouth of its river a couple of miles up the coast. With its sheltered anchorage gone, the town's attractiveness as a port disappeared, its sea defences were neglected, and coastal erosion clawed away at the land.

Deposition, the opposite of erosion, can also take its toll. Deposition is the process by which sediments (clay, sand, silt, shingle) are added to the land by such factors as river flow, winds and maritime currents. It was this that did for Sandwich, in eastern Kent. Today Sandwich is known for its golf courses, but in the Middle Ages it was an important port, one of the famous five Cinque Ports. But its river silted up, the coastline shifted eastward into the North Sea, and the town is now stranded two miles inland.

More often humans have made deliberate efforts to make land where once there was water. The draining of the Fens from the early seventeenth century by imported Dutch engineers is just one example. The islands of Canvey and Wallasea were similarly drained, and surrounded with sea walls. In the case of Wallasea much of the island has now been purchased by the RSPB, who have undertaken the task of reversing the work of the land reclaimers. Breaches have deliberately been made in the island's defences to let the sea in. The east of the island is now tidal wetland, supporting a considerably greater diversity of wildlife than the arable monocultures that dominate much of the English countryside.

Many experts now believe that erecting or maintaining hard coastal defences such as sea walls and groynes in one place only shifts the problem elsewhere. There has been a preference for softer approaches, such as stabilising dunes by encouraging marram grass and similar species that help to bind the sand. In sparsely populated places – such as Wallasea – the policy is 'managed retreat', whereby the sea is allowed to have its way with the land. There has been a growing realisation that human efforts to control the sea are doomed to failure, and that meddling only makes matters worse. A stark reminder of the transience of human endeavour came in December 2013, when a storm surge driving down the North Sea cut the long, thin sandbar linking Spurn Head, at the mouth of the Humber, from the rest of Yorkshire, at least at high tide. The road to the point was destroyed, and now it is only possible to reach Spurn Head on foot, at low tide. Spurn became known as 'England's newest island'. The policy now in many places is to let nature take its course.

The encroachment of the sea upon the land is set to increase as the earth warms, largely owing to human activity. This is not just a local issue. If sea levels rise, they do so globally. We may strive to protect areas of denser population with concrete walls, but in the end it is up to humans to address the actual causes of the danger. The foothold of humans on this earth is a tenuous one. In the face of the indifference of nature, of a planet heedless of our purposes, or even our presence, we come up with constructions such as the nation-state, as if such constructions will save us.

One such nation-state, the Republic of Maldives, comprises a chain of atolls in the Indian Ocean. On average the country is less than five feet above sea level. In 2007 the Intergovernmental Panel on Climate Change reported that, if things do not change, by 2100 most of the 200 inhabited islands in the Maldives will have to be abandoned. There is even talk of purchasing land in India, Sri Lanka and Australia for the displaced populations. Many of England's islands, from those in the estuaries of the southeast to the Isles of Furness in the northwest, are little higher than the Maldives.

The British – particularly the residents of Little England – talk a lot about 'taking control' of their borders. They believe they have an inalienable right over a fixed area of the earth's surface. But there is both folly and vanity in staking a claim over a parcel of land whose frontiers are always in flux. The sea constantly shifts the boundaries. As they have always done, seals pull up on the beaches to pup, then swim off again into a wider, less bounded world; birds fly freely overhead; seeds blow in the wind. Humans too have always moved. There were people living in what is now England eons before the English, before the Romans, before the Celts. And the English today are a blend of all sorts of peoples who have come from elsewhere, not only Romans and Celts, Angles and Saxons, but Vikings and Normans, Hollanders and Huguenots, Jews and Jamaicans, Eastern Europeans and Southern Asians, flows of different cultures, mingling and entwining, enriching each other like ocean currents. No man, nor woman either, is an island.

Index